Quality Control and Artefacts in Magnetic Resonance Imaging (Update of IPEM Report 80)

IPEM Report Number 112

A Report produced by the Institute of Physics and Engineering in Medicine

Editors
Donald McRobbie, Adelaide and London
Scott Semple, Edinburgh

Authors
Anna Barnes, London
Geoff Charles-Edwards, London
Glyn Coutts, Manchester
Peter Gatehouse, London
Martin Graves, Cambridge
Katherine Lymer, Edinburgh
Ian Marshall, Edinburgh
Aaron McCann, Belfast
Donald McRobbie, Adelaide and London
Stephen Nicholas, Dundee
Annie Papadaki, London
Andrew Patterson, Cambridge
Scott Semple, Edinburgh
John Thornton, London
Nick Weir, Edinburgh
Daniel Wilson, Leeds
Stephanie Withey, Birmingham
Gavin Wright, Leeds

Contributors

Anna Barnes
University College London Hospitals NHS Foundation Trust, London

Stephen Nicholas
Ninewells Hospital and Medical School, Dundee

Geoff Charles-Edwards
Guy's & St Thomas' NHS Foundation Trust, London

Annie Papadaki
University College London Hospitals NHS Foundation Trust, London

Glyn Coutts
The Christie NHS Foundation Trust, Manchester

Andrew Patterson
Cambridge University Hospitals NHS Foundation Trust, Cambridge

Peter Gatehouse
Royal Brompton and Harefield NHS Trust & Imperial College, London

Scott Semple
Clinical Research Imaging Centre, University of Edinburgh, Edinburgh

Martin Graves
Cambridge University Hospitals NHS Foundation Trust, Cambridge

John Thornton
University College London Hospitals NHS Foundation Trust, London

Katherine Lymer
SFC Brain Imaging Research Centre, University of Edinburgh, Edinburgh

Nick Weir
Royal Infirmary of Edinburgh, Edinburgh

Ian Marshall
University of Edinburgh, Edinburgh

Daniel Wilson
Leeds Teaching Hospitals NHS Trust, Leeds

Aaron McCann
Belfast Health & Social Care Trust, Belfast

Stephanie Withey
University Hospitals Birmingham NHS Foundation Trust, Birmingham

Donald McRobbie
Division of Medical Imaging, Flinders Medical Centre, Adelaide, Australia and Department of Surgery, Imperial College Faculty of Medicine, London

Gavin Wright
Leeds Teaching Hospitals NHS Trust, Leeds

The Editors and authors would like to acknowledge the valuable contribution of Paul Tofts in suggesting additional material and references throughout this report.

Fairmount House, 230 Tadcaster Road
York YO24 1ES
ISBN 978-1-903613-61-0

Published by the Institute of Physics and Engineering in Medicine
Fairmount House, 230 Tadcaster Road, York YO24 1ES

Prepared by:
The Charlesworth Group, Wakefield, UK. www.charlesworth-group.com

Contents

Part I –
Quality Control for MRI

1 Introduction

Donald W. McRobbie

1.1 General remarks

Since the original publication of IPEM Report 80 (Lerski *et al.*, 1998), there have been a significant number of developments in clinical magnetic resonance imaging (MRI). Many of these are technology-driven, for example, the migration to higher static field strengths, or the development of multi-channel radiofrequency (RF) systems and parallel imaging techniques. There have also been significant developments in many clinical areas, particularly in cardiac MRI, MR perfusion, functional (brain activation) spectroscopy, MR in radiotherapy planning and real-time guidance for surgical interventions and biopsies. Additionally, there has been a recent resurgence in quantitative MRI, for example, in permeability imaging, relaxometry and quantitative techniques using calculated parameters, including fibre tractography, dynamic breast imaging, volumetric and quantitative morphological studies. MRI is now an established biomarker in pharmaceutical development research where accuracy and precision of image-derived parameters are of paramount importance.

In each of the above areas, there is a need for accurate, reliable imaging and for the determining and assuring of image quality, parameter accuracy and patient safety. Therefore the need for Quality Assurance (QA) systems and Quality Control (QC) activities has never been greater. In addition to all these new and emerging areas, the traditional aspects of QC in sequence evaluation, scanner performance assessment and acceptance testing are still vitally important to the provision of efficient and effective imaging.

Primarily, this publication addresses general quality control for MRI, but with clinically relevant sections devoted to particular specialist areas that are dependent upon measurement accuracy and precision. Image artefacts are an important component of image quality, often rendering images undiagnostic, and are considered in detail.

1.2 Review and background

1.2.1 Standards

Standards for MR image quality evaluation have previously been published by the North American Equipment Manufacturers Association (NEMA, 2008a,b,c,d,e, 2010a,b), the American Association of Physicists in Medicine (Price *et al.*, 1990; AAPM, 2010), the International Electrotechnical Commission (IEC, 2007) and the American College of Radiology (ACR, 2005). Of these, the ACR series of standards remains the most comprehensive, forming part of an accreditation scheme which includes the evaluation of clinical scans in addition to phantom-based tests. The

NEMA standards define suitable measurement methods for various image quality parameters such as signal-to-noise ratio (SNR), slice thickness and spatial resolution but do not prescribe specific test objects or pulse sequences. A similar approach is taken by the IEC. These methods will be reviewed below in specific sections.

The Eurospin Test System (European Communities Research Project (COMAC BME II 2.3), 1998; Lerski *et al.*, 1998; Lerski and de Certaines, 1993) was developed as part of a project from the early days of clinical MRI to characterise relaxation times but may prove cumbersome for evaluating modern scanner performance due to their limitations (principally designed to be single-slice, parameter-specific objects). The earlier IPEM Report 80 focused specifically on the use of the Eurospin Test Objects. The same European group has also published quality control methods for spectroscopy (European Communities Research Project, 1995).

1.3 The Quality Cycle, QA and QC

Quality Assurance is the overarching system or process to ensure the ongoing delivery, maintenance and improvement in the quality of a service or product. The Medicines and Healthcare products Regulatory Authority's (MHRA) Device Bulletin "Safety Guidelines for Magnetic Resonance Imaging Equipment in Clinical Use" (MHRA, 2014) gives advice on equipment management, starting with the formation of a project team, identification of specialist advisors, use of published data, technical questionnaires, drawing up of specifications, acceptance testing, ongoing QC programmes, maintenance and corrective actions, and finally planning for replacement. Together these constitute the quality cycle of the equipment. The QC component consists of the technical procedures and measurements necessary to demonstrate compliance with standards or specifications. QC is the focus of this report.

1.3.1 Acceptance testing

Acceptance testing is a specialised QC activity which forms an essential part of the quality cycle, forming a link between the equipment specified and its ongoing service life. It provides the opportunity to confirm that the equipment installed is performing in such a manner as to be clinically acceptable and also provides a baseline for ongoing QA programmes. Responsibility for the performing of acceptance tests rests with the host institution. The warranty period is usually deemed to begin with the signing of the acceptance form.

Acceptance tests should include measurements of signal parameters (SNR, uniformity), geometric parameters (spatial resolution, linearity, distortion), slice and relaxation parameters (slice position, slice thickness, contrast, contrast-to-noise ratio (CNR)), artefacts (ghosting, RF lines, water–fat suppression/selection and any other artefacts). SNR should be evaluated for every coil under appropriate loading conditions. An audit of acceptance testing (McRobbie and Quest, 2002) found that SNR and geometric linearity were the parameters most prone to fall below expected standards.

Whilst the acceptance test will focus upon the measurement of image quality parameters, it is unlikely that the MR vendor will specify performance in these terms. Rather the vendor's specifications will be in terms of the engineering performance of the system, e.g. static field (B_0) homogeneity, temporal stability and fringe field; gradient amplitude, slew rate and linearity; RF peak amplitude, power and stability. Some of these are addressed in AAPM Report 100 (AAPM, 2010). It would be prudent to define the customer's image quality expectations before the procurement process. An example of these is given by McRobbie and Quest (2002).

Non-imaging tests should also be part of acceptance testing, including the verification of the fringe field contours, acoustic noise and electrical safety, although these may be performed by non-MRI specialists. For a fuller description of these, see IPEM Report 98 "Guidance on Electromagnetic Fields and Electromagnetic Compatibility" (McRobbie, 2010) and the relevant NEMA and IEC standards and AAPM Report 100.

1.3.2 Routine QA/QC

Serial QC may be used for monitoring scanner performance over time. This highlights drift in performance, enabling successful implementation of preventative measures before they reach clinical significance and providing evidence to support an equipment replacement business case. Quality is also a key principle of clinical governance (Koller *et al.*, 2006).

Imaging parameters which may usefully be monitored routinely include: SNR, level of artefacts and geometric scaling. Firbank *et al.* (2000) analysed QC data performed daily, weekly and monthly over 1 year, concluding that SNR, geometric distortion and RF amplifier voltage should be measured weekly. Chen *et al.* (2004) and Mulkern *et al.* (2008) have evaluated the use of the ACR phantom in both single- and multi-site situations. Ihalainen *et al.* (2004) carried out similar work using the Eurospin Test Objects. In a multicentre trial, Colombo *et al.* (2004) concluded that weekly QC measurements were adequate to detect variations in system performance.

Routine QC/QA specified by the equipment manufacturer usually involves the use of manufacturer's phantoms and software for automatic analysis and reporting. Whilst this is very simple to apply, often the level of reporting to the user, e.g. a simple in or out of tolerance, is provided, and such tolerances are system-specific. Also the majority of manufacturer's routine QC concentrates on the stability of mechanical components and electronic circuits but image quality may not be optimised despite complying with manufacturer's specification (Chen *et al.*, 2004). Consequently, there is a need to supplement these with user-defined QC procedures. The MHRA strongly recommends that clinical institutions have written policies for independent MRI equipment image quality testing, particularly the monitoring of signal and geometric parameters (MHRA, 2014).

1.3.3 Specialist QC

Specialist QC may be required on a regular or per patient basis where particular, often quantitative, clinical investigations are to be carried out. Examples include performing clinical fMRI (Friedman and Glover, 2006), where short-term scanner stability is of great importance or serial tumour response measurements, where long-term stability may be very important (Velthuizen *et al.*, 1998; Gedamu *et al.*, 2008; Cormack, 2008; Tofts and Collins, 2011). Other areas where specialist QC is essential include MR treatment planning for radiotherapy (Chen *et al.*, 2006) and radiosurgery (Poetker *et al.*, 2005), MR-guided high intensity focused ultrasound (Wu and Felmlee, 2002; Gorny *et al.*, 2006) where geometric accuracy (Baldwin *et al.*, 2007) is paramount, multi-modality registration (Isambert *et al.*, 2008), quantitative studies of relaxation times (Ciarmiello *et al.*, 2001), for example, the assessment of organ iron overload using T_2 or T_2*, spectroscopy (van der Graaf *et al.*, 2008; Gu *et al.*, 2008), diffusion MR (Delakis *et al.*, 2004; Lavdas *et al.*, 2013), cardiac MR (Woodward *et al.*, 2006) and quantitative flow measurement (Marchand *et al.*, 2002). This list is not exhaustive.

1.4 Requirements for QA/QC

1.4.1 Test objects

Test objects are usually used for measuring image quality parameters. The technical requirements of test objects are considered in Chapter 2. Example test objects include those from the ACR (Figure 1.1a) and the Eurospin Test Objects (Figure 1.1b). Test objects may be multi-parameter, providing a wide range of image quality metrics as in the ACR object, or cover a defined range of parameters, often related, as in Eurospin. Multi-parameter objects are preferable. Test objects may be commercial or bespoke. Particular specialist applications may require specialist objects, for example, in flow measurement (Fahrig *et al.*, 1999), diffusion imaging (Reischauer *et al.*, 2009; Lavdas *et al.*, 2013) or fMRI (Olsrud *et al.*, 2007).

The design of the test object may be such that evaluation is subjective or objective. Subjective methods are preferable for a "quick check" but lack the trend prediction power afforded by serial objective measurements. Objective measurements require further image processing or analysis that may have to be performed off-line. Figure 1.2 shows image examples of both objective and subjective image evaluation features in the ACR phantom and a bespoke multi-parameter, multi-slice phantom (McRobbie, 1997).

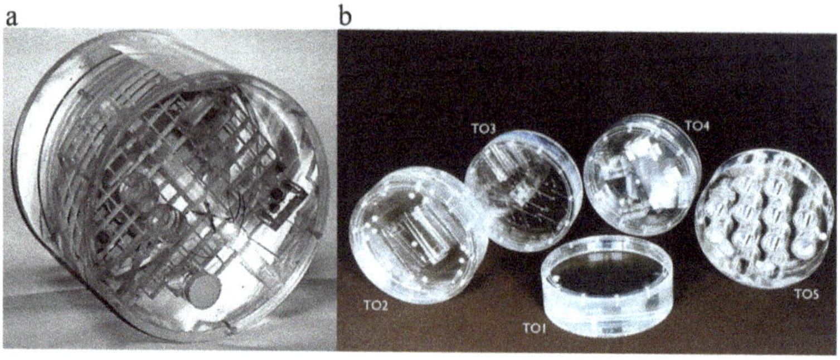

Figure 1.1 (a) ACR phantom (courtesy of the American College of Radiology); (b) Eurospin Test System (courtesy of Diagnostic Sonar Ltd)

1.4.2 Sequences

The appropriate choice of MR pulse sequence and acquisition parameters is of critical importance to the interpretation of QC data. A significant component of any system for assessing image quality must be to define the testing conditions. Parameters such as spatial resolution and SNR cannot be properly interpreted without knowledge of the sequence parameters. For example, changing the field-of-view (FOV) without changing the acquisition matrix will alter the spatial resolution, and changing the slice thickness or pixel dimension will alter the SNR. For a review of parameter dependence upon image quality, see McRobbie *et al.* (2007). Other parameters are more prevalent for specific sequences, for example, echo planar imaging (EPI) is more likely to produce

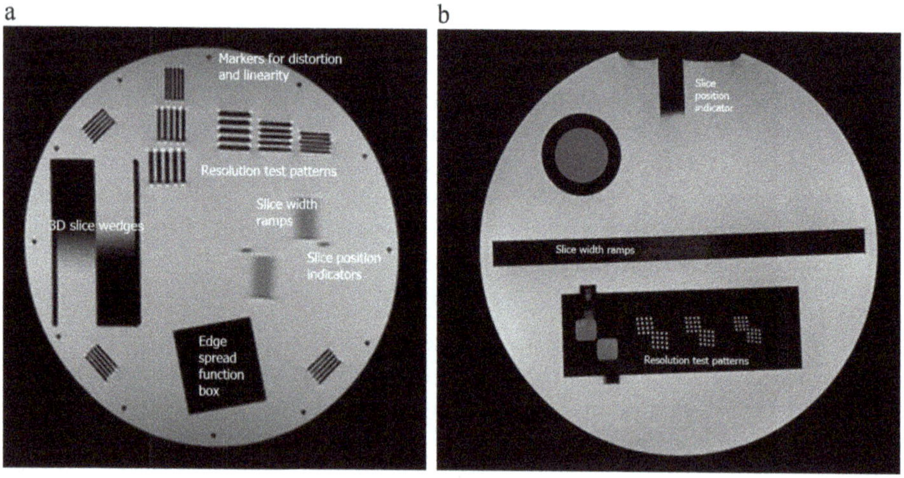

Figure 1.2 Examples of objective and subjective assessment of image quality – in (a) resolution and slice position may be determined objectively, in (b) the assessment is determined subjectively

Table 1.1 The interdependence of image quality parameters. The evaluation of each parameter listed in the first column is affected by other parameters ticked across the rows

Primary parameter	Parameters affecting primary parameter									
	SNR	Uniformity	Geometric linearity	Geometric distortion	Resolution	Slice thickness	Slice position	Ghosting	Contrast	CNR
SNR[a]		✓	✓		✓	✓	✓	✓	✓	✓
Uniformity[a]	✓					✓		✓	✓	
Geometric linearity[b]				✓	✓		✓			
Geometric distortion[b]			✓				✓			
Resolution[b]			✓							
Slice thickness[c]		✓	✓				✓			
Slice position[c]			✓	✓						
Ghosting[d]	✓	✓								
Contrast[c]		✓				✓				✓
CNR[c]	✓	✓			✓	✓			✓	

[a]See Chapter 3 for SNR (signal-to-noise ratio) and uniformity measurement.
[b]See Chapter 4 for geometric parameter definitions.
[c]See Chapter 5 for relaxation parameters. CNR is contrast-to-noise ratio.
[d]See Chapter 6 for ghosting artefacts.

phase ghosts and geometric distortion than most other conventional sequences. These issues will be addressed in each section below, where relevant. Table 1.1 provides an indication of the interdependence of various image quality parameters, with more details and explanation given in the relevant chapters.

1.4.3 Analysis software

For the evaluation of quantitative parameters, some rudimentary image processing or evaluation may be required. These will include pixel value statistics (mean, standard deviation, median, area) from regions of interest (ROI), linear and angular measurements, and line profiles. The scanner software may be limited in what can be measured, for example, limited ROI shapes may be available and the ability to average several line profiles, or to export data in text files. The ubiquitous use of the DICOM image format (NEMA, 2011) means that transferring images to independent workstations or computers is no longer an issue. Commercial software such as Analyze™ (AnalyzeDirect Inc., Overland Park, Kansas, USA) or shareware software such as Image-J or Osirix may be used for analysis. Some authors have developed automated software for QA (Chen *et al.*, 2004; Colombo *et al.*, 2004; Sun *et al.*, 2015).

2 Principles of test object construction

Scott Semple

2.1 General remarks

Robust test objects (commonly referred to as 'phantoms') are essential to ensure that routine quality assurance is performed with precision and reproducibility. Phantoms must be constructed of materials that have suitable levels of radiofrequency (RF) absorption, magnetic susceptibility effects, ferromagnetic properties, electrical conductivity, dielectric constant and MR relaxation times. Phantoms may be obtained from scanner manufacturers, third party companies, or constructed by medical physics departments if suitable materials and workshop facilities are available.

2.2 Construction materials

2.2.1 Susceptibility

When a material is placed within an external magnetic field H (A m^{-1}), the internal magnetisation M (A m^{-1}) results from interactions with H according to

$$M = \chi H \tag{2.1}$$

where χ is the magnetic susceptibility of the material and is dimensionless. When two materials of differing magnetic susceptibilities are in close proximity, local distortions in the static magnetic field B_0 are observed. These static field inhomogeneities cause dephasing of nearby spins, resulting in localised image artefacts (signal loss and image distortion) around the susceptibility boundary.

This effect commonly occurs in vivo such as between tissue and air interfaces (for example, around the paranasal sinuses). Between water and pure oxygen, a local field difference of 1.33 parts per million (ppm) is observed. Air bubbles within phantoms may therefore cause local static field artefacts as shown in Figure 2.1. Interfaces between water and Perspex result in a field difference of only 0.004 ppm. Static field and other artefacts are discussed in more detail in Part II (Chapter 8).

Any metal (ferromagnetic or not) causes severe variations in local magnetic fields, causing significant static field image artefacts. All metal should therefore be avoided in construction of phantoms to avoid resultant image artefacts.

2.2.2 Ferromagnetism

Phantoms must be constructed entirely from non-ferromagnetic materials. Any ferromagnetic items will induce significant image artefacts including signal drop-out, as well as being a potential safety hazard (missile effect).

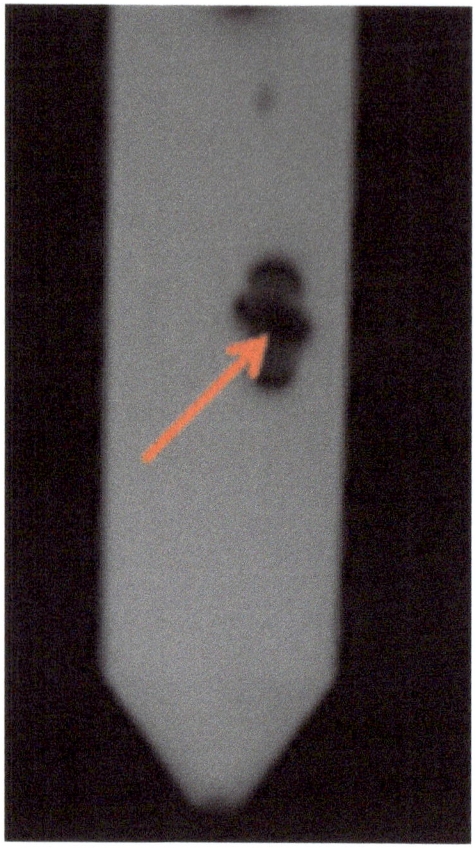

Figure 2.1 Gradient echo (TE=10 ms, slice width=5 mm) of 50 mL agar gel test object demonstrating significant static field artefact around a small air bubble within the gel phantom (red arrow)

2.2.3 RF absorption

When constructing a phantom, materials should be used which have a RF absorption similar to that of tissues. If a material with significantly different RF absorption to tissue was used, the resultant difference in signal intensity and/or gain settings applied by the scanner could lead to misleading results. For example, a material with RF absorption properties that are significantly greater than tissue may result in a scanner receiver gain setting significantly greater than used in imaging humans. Most common materials used for the construction of phantoms have similar RF absorption rates to tissues, mostly governed by the electric field distribution within the material (magnetic permeability of H_2O, air, and most filling materials is approximately equal).

2.2.4 Summary

For phantom construction, Perspex and/or glass are commonly used for the main construction, and pieces are joined with either plastic screws and/or glue. No metal should be used in the construction.

2.3 Filling materials

The material used to fill phantoms should have properties that mimic the MRI properties of tissue. Phantoms are commonly filled with gels, oil or water, which are then doped with paramagnetic salts such as nickel or copper sulphate. These salts are added to bring the MR relaxation times of the phantom in line with the range found in tissue in vivo.

At higher field strengths (3 T and above), standing wave effects may necessitate the use of oil-filled phantoms. The B_1 transmit field has a finite wavelength, which can be described in vacuum, or in air, by

$$\lambda_{VACUUM} = c/f \tag{2.2}$$

where f is the frequency and c is the speed of light. At 1.5 T, the wavelength in vacuum or air is approximately 4.7 m. However, in a material with relative permittivity ε_R and relative permeability μ_R

$$\lambda_{MATERIAL} = \lambda_{VACUUM} \Big/ \sqrt{\varepsilon_R \mu_R} \tag{2.3}$$

Water has a very low magnetic susceptibility such that μ_R equals 1.0, but has a substantial ε_R of around 80, therefore the wavelength is only 0.52 m, and of the same order as the range of fields-of-view (up to 50 cm).

However, the distance between nodes and anti-nodes of the RF standing wave pattern is actually $\lambda/4$. At anti-nodes, the transmit field is negligible such that areas of low signal will be seen. At 1.5 T, $\lambda/4$ for water is only 13 cm. In vivo, tissue boundaries and membranes break up standing waves and prevent them forming, however, in water-filled phantoms, anti-nodes in the standing waves may give a false indication of a poor signal uniformity. Since oil has a ε_R in the range of 3–5, using oil-filled phantoms can overcome this problem (Tofts, 1990).

2.3.1 Proton density

If aqueous solutions are used then the proton density of the phantom will be effectively 100%, which is 20–30% higher than most tissues in vivo. Imaging of purely aqueous solutions may therefore lead to overloading of the receiver gain of the system if used at standard settings. Construction of the phantom using gels rather than aqueous solutions may be used to lower the proton density of the solution to a level similar to tissues, or paramagnetic salts added to modify the relaxation times and hence the detected MR signal to the appropriate range.

2.3.2 Relaxation properties

Although phantoms should ideally mimic the relaxation properties of tissues, there are a range of T_1 and T_2 values in various tissues in vivo (de Certaines *et al.*, 1993; Henriksen *et al.*, 1993), and T_1 increases with increasing field strength whilst T_2 is generally constant, decreasing slightly at 3 T as shown in Table 2.1. Therefore, phantoms may need to be constructed for specific applications (i.e. to mimic certain tissue relaxation times) at specific field strengths. For example, a head phantom might aim to mimic the relaxation properties of grey matter, white matter, and CSF.

For water-based phantoms, paramagnetic salts are used to reduce the very long relaxation times of water to values approximating those of tissues. A list of commonly used suitable paramagnetic salts used in phantom construction is shown in Table 2.2. Other salts are available. The combination and strength of these salts are dictated by the desired relaxation properties.

Table 2.1 Selection of T_1 and T_2 values for tissues at 0.5, 1.5 and 3.0 T. All values measured in vivo from human tissues (reproduced from McRobbie *et al.*, 2007)

Tissue	**T_1 (ms)**			**T_2 (ms)**		
	0.5 T	**1.5 T**	**3.0 T**	**0.5 T**	**1.5 T**	**3.0 T**
White matter	520[f]	560[a]	832[i]	107[b]	82[c]	110[i]
Grey matter	780[f]	1100[a]	1331[i]	110[b]	92[c]	80[i]
CSF	–	2060[e]	3700	–	–	–
Muscle	560[g]	1075[d]	898[h]	34[g]	33[g]	29[h]
Fat	192[b]	200[b]	382[h]	108[b]	–	68[h]
Liver	395[b]	570[e]	809[h]	96[b]	–	34[h]
Spleen	760[b]	1025[e]	1328[h]	140[b]	–	61[h]

[a]Steinhoff *et al.*, 2001;
[b]Bottomley *et al.*, 1984;
[c]Pfefferbaum *et al.*, 1999;
[d]Venkatesan *et al.*, 1998;
[e]Bluml *et al.*, 1993;
[f]Imran *et al.*, 1999;
[g]de Certaines *et al.*, 1993;
[h]de Bazelaire *et al.*, 2004;
[i]Wanspura *et al.*, 1999.

Table 2.2 Paramagnetic salt relaxation times (adapted from McRobbie *et al.*, 2007)

		Ratio of T_1/T_2	**Relaxivity R_1 (s^{-1} $mmol^{-1}$)**	
			0.07 T	**1.4 T**
Copper sulphate	$CuSO_4$	1.1	1.7	0.5
Manganese chloride	$MnCl_2$	2–10[a]	13	7
Gadolinium chloride	$GdCl_2$	1.1	22.1	10
Nickel chloride	$NiCl_2$	1.1	0.6	0.65
Normal tissue		4.5		

[a]$MnCl_2$ T_1/T_2 ratio varies with B_0.

In water-based solution doped with paramagnetic salt, the T_1 dependence on paramagnetic concentration is given by

$$R_1{}^{SOLUTION} = \alpha.C + R_1{}^{WATER} \tag{2.4}$$

where R_1 is the longitudinal relaxation rate ($1/T_1$ in s^{-1}), C is the molar concentration of the salt (mol L^{-1}), and α is the longitudinal relaxivity of the paramagnetic molecule (s^{-1} mol^{-1} L). Similarly

$$R_2{}^{SOLUTION} = \beta.C + R_2{}^{WATER} \tag{2.5}$$

where β is the transverse relaxivity of the paramagnetic molecule.

The ratio of T_1/T_2 for common salts is shown in Table 2.2. Construction of gel phantoms offers the advantage of allowing T_1 and T_2 values to be adjusted independently. T_1 values of gel phantoms depend on the paramagnetic ion concentration, whilst T_2 is usually dependent on the gel concentration; Carrageenan gel is an exception where concentration has little influence on the T_2 relaxation time (Yoshimura *et al.*, 2003; Hattori *et al.*, 2013).

T_1 relaxation times are also dependent on temperature. Most T_1 relaxation times for paramagnetic salts increase with temperature (particularly so for copper sulphate), other than nickel salts, which show a slight decrease with temperature. However, over the range of normal room temperatures, the relaxation times of most materials can be thought of as relatively constant. For accurate T_1 measurements (particularly when measured in longitudinal studies or for regular QC), an accurate assessment of bore temperature should be made.

2.3.3 Electrical properties

When applying RF fields to conducting objects, RF eddy currents may be set up near the surface. This is known as the "skin" effect (Lorrain and Corson, 1970).

In vivo, tissue boundaries, membranes and air spaces tend to break up eddy currents and reduce this effect, however at higher field strengths (⩾1.5 T), saline phantoms such as aqueous NaCl phantoms, may give a very non-uniform response. A large saline phantom may therefore give a false indication of poor signal uniformity. One solution is to use a loading annulus that is doped with NaCl around the phantom to break up the eddy current effect.

2.3.4 Summary

Phantoms should be filled with materials that mimic the magnetic properties of tissue as closely as possible. Gels or paramagnetic-doped water are commonly used for this purpose.

2.4 Practical issues

2.4.1 Designing your own

When designing a phantom, the architecture of the coil being tested should be taken into account. The phantom should be of sufficient size and shape to adequately fill the coil and mimic the area of the body normally imaged using that coil. The phantom should fill a standard FOV commonly used in a protocol for the coil. Therefore, a body or torso coil phantom will be significantly larger than a head or knee coil phantom.

Phantom design may also mimic a degree of function of the tissue or body part being imitated. For example, several types of semi-solid breast phantoms have been constructed with internal gel structures of appropriate relaxation values to mimic both tissues and lesions for biopsy and allow QC of breast compression and biopsy devices (Freed *et al.*, 2011). Some of these phantoms have also been specifically designed for use across different imaging modalities commonly used in breast imaging, and as such, the component parts/materials must also be designed to be ultrasound- or radio-opaque (Freed *et al.*, 2011; Tang *et al.*, 2008).

2.4.2 Differences between phantoms and human subjects

It should be noted that phantoms only mimic some of the basic properties of part of human subjects, for example, in terms of coil loading, RF absorption, relaxation times, etc. Imaging of human subjects is significantly more challenging and can result in different problems. For example, human motion within the scanner can be a significant factor affecting image quality. Any imaging within the thorax is prone to artefacts caused by breathing, cardiac, or intestinal motion. Gross human subject movement within the scanner may also increasingly be a problem if long imaging times are required.

Within a normal FOV in human subjects, there are commonly a wide variety of tissue types with a range of relaxation parameters. Attempting to mimic both the geometry and relaxation values of the range of tissues commonly present in a single

FOV is impossible. Most phantoms only attempt to test coil characteristics rather than mimic human subjects. Some phantoms that mimic specific properties such as flow or cardiac movement have been implemented for specific studies (Khan and de Wilde, 2003). The entire list of phantoms designed to mimic specific biological properties or function is almost endless, and beyond the scope of this text. Some common examples are covered in later chapters, such as T_1, and T_2 relaxation phantoms (Chapter 6), and phantoms suitable for diffusion imaging (Chapter 7).

The loading effects of human subjects within the scanner will not be adequately modelled by phantoms, with the average human body significantly distorting the homogeneity of the main B_0 field. This distortion of the B_1 transmit field in the presence of the human body is also problematic when imaging at higher field strengths, particularly when performing breast imaging, where the subject is commonly imaged prone above a raised breast coil. The asymmetric loading of the system causes significant B_1 distortion, resulting in spatial variations in flip angle across the FOV (Kuhl *et al.*, 2007). This effect is typically not as prominent when imaging phantoms.

RF eddy currents set up by the applied B_1 field near the surface of conducting objects, atypical RF absorption and standing waves in phantoms may differ significantly from tissues, giving a very non-linear signal response, which does not accurately represent the response from a human subject.

2.4.3 Mechanical construction

When constructing phantoms, the earlier advice on use of non-ferromagnetic materials applies. Phantom components should be precision constructed and screwed tight with non-ferromagnetic screws and glue-sealed so as to be watertight for obvious reasons. Commercial phantoms for measuring uniformity and SNR are commonly spherical in shape and allow imaging in three orthogonal planes without removal and repositioning in the scanner. For user-made phantoms, a thick disc shaped phantom is commonly used instead of a sphere as these are comparatively easy to construct (to image in different planes, such a phantom would need to be rotated). For testing spatial distortion, slice profile and other geometric parameters, Perspex wedges, block or rods may be glued in place within the main body of the phantom (disc shaped phantoms are also usually constructed for these purposes). Phantoms should be constructed with a resealable watertight aperture for filling so as to allow periodic 'top-up' of filling material. Any phantoms constructed in-house would of course need to be validated externally or against a pre-validated gold-standard in some manner.

2.4.4 Making filling materials: fluids and gels

It is useful to adjust the pH of phantoms to approximately 4.0 by adding a very small amount of sulphuric or hydrochloric acid to prevent precipitation of Cu ions as hydroxyl which would lead to an increase in T_1 of the solution. A preservative agent may also be added to prevent bacterial or fungal growth. This often gives phantoms the appearance of being slightly 'cloudy'.

2.4.5 Filling of phantoms

Once the filling material has been mixed, phantoms should be filled with a minimal air gap around the filling aperture. Too large an air gap will result in significant inhomogeneity artefacts around the fluid/air interface. A small air gap around the filling aperture may allow for temperature-dependent expansion and contraction of the filling material.

Some loss of filling solution due to oxidation and lack of water tightness is expected. Phantoms should therefore be periodically topped up with filling material. It is therefore pertinent to keep a store of additional top-up filling material from the initial batch of filling material. It is prudent that if there is no remaining filling material from the initial batch, then the phantom is completely refilled from a new batch. Mixing older and newer batches of filling material is not advised, particularly when the phantoms are to be used for T_1 or T_2 measurements.

2.4.6 Care of phantoms

Phantoms are commonly fluid filled, glass or Perspex encased objects. As such, they are usually quite fragile. Construction of safety storage for each phantom is advised (padded 'flight cases' are commonly used for this purpose).

To avoid 'swirling' effects (and the concurrent artefacts) when imaging phantoms, it is best to allow aqueous phantoms to settle after moving them into imaging position within the scanner. Common clinical practice is to load the daily QC phantoms into the scanner at the end of the working day for the following morning's pre-scan QC checks.

Periodic QC of phantoms should be performed, such as checking fluid levels, taking regular T_1 and T_2 measurements, measuring pH, and checking for any leaks or structural defects (such as any Perspex wedges or blocks coming loose within the phantom itself).

2.4.7 Choosing a commercial phantom

Most commercial scanners come with a set of vendor supplied phantoms which are routinely used by vendor service engineers. It is common to be provided with a range of uniformity phantoms suitable for each coil provided, and several resolution/slice profile phantoms. Additional phantoms may be provided with individual third-party coils, for example, interventional breast MR coils or wrist coils which are supplied with a phantom of specific size, shape and appropriate relaxation properties to enable suitable QC measurements to be performed for the coil.

Several third-party manufacturers also fabricate specialist phantoms designed to mimic either anatomical detail, or physiological function, for example, cardiac-specific MR phantoms designed to mimic anatomy, function and/or flow. In all cases when purchasing commercial phantoms, the user must ensure that the phantom functions as required. Some commercial phantoms are designed with one very specific function in mind and may not be suitable for a range of QC applications. Some commercial phantoms may be expensive, so great care must be taken to investigate phantom function before purchase.

A comprehensive list of all commercially available phantoms is beyond the scope of this text (new phantoms are being developed and commercialised almost constantly).

There are several multi-purpose phantoms that are commonly used in most MRI centres. These types of multi-purpose phantoms are an ideal option to address most QC requirements. These would include, but are not limited to, the American College of Radiology (ACR) MR accreditation phantom (capable of assessing geometric distortion, spatial resolution, slice thickness/position, interslice gap, bandwidth, uniformity SNR), the Fluke Biomedical multi-purpose phantom (capable of assessing slice thickness, slice orientation, interslice gap, homogeneity, spatial resolution, T_1 and T_2 values with other assessments being possible using Fluke Biomedical's other commercially available phantoms) and Diagnostic Sonar's Eurospin phantom set (the set is capable of assessing uniformity, slice width and position, spatial resolution, contrast and relaxation times). The use of these phantoms for specific QC assessments is discussed in more detail in later chapters. These are some useful examples only and this list is not meant to be comprehensive (the authors do not specifically endorse the use of any commercial phantom). Each site should evaluate their own QC requirements carefully before purchase of any commercial phantom.

3 Signal parameters

Stephen Nicholas

3.1 Introduction

In MRI, a signal is detected as an induced voltage in a radiofrequency (RF) coil. This signal contains two components: the first arises from the precession of magnetisation vectors in the object being imaged that have been tipped into the transverse plane by an RF pulse. The second arises from random electromagnetic fluctuations in the object and imaging system. The quality of an image will depend on the relative contribution of these two components, known as the signal-to-noise ratio (SNR) and also the intrinsic variability of the signal in space (uniformity) and whether the signal is assigned the correct spatial position in the final image (ghosting).

This chapter introduces these three concepts: SNR, uniformity and ghosting, and describes how they can be quantified and their impact on image quality.

3.2 Signal-to-noise ratio (SNR)

The SNR is a measure of how the signal and hence the pixel intensity in the image is affected by random fluctuations. These random fluctuations, referred to as noise, arise from several places, including electronic noise in the RF coil and receiver system as well as fluctuations in the magnetic field from random ionic currents in the object being imaged. A more detailed description of signal and noise in MR imaging can be found in Hoult and Richards (1976), Hoult and Lauterbur (1979), Redpath (1998) or Haacke *et al.* (1999).

SNR is an important parameter as detail and contrast in clinical MR images with low SNR are often obscured, and this could lead to a misdiagnosis. SNR depends on many different factors. However, it represents a good single measure of image quality that can be easily monitored over time. Most hardware or software related faults will be reflected in changes in SNR values, implying that regular SNR measures are a vital part of any MR system QA programme. The main factors affecting SNR are described in detail below.

3.2.1 Sequence type and parameters

The choice of sequence type and parameters (e.g. repetition time (TR), echo time (TE), flip angle, field of view (FOV), matrix size, slice thickness, receiver bandwidth, etc.) can have a major effect on measured SNR and it is therefore necessary to use a standardised protocol with specific parameters (see Section 3.3.4). In some cases, variations in sequence parameters may be accounted for by applying correction factors. Normalised SNR may also be used to compare measurements between different scanners (McRobbie, 1997).

3.2.2 Phantom composition and relaxation times

The signal in any SNR measurement will be affected by the relaxation times of the phantom material. Aqueous solutions of nickel chloride or copper sulphate are often used to give suitable phantom T_1 and T_2 values (see Chapter 2). Diluting the phantom solution will lengthen T_1 and T_2, which may change the signal. At high field strengths, oil is used to fill phantoms to minimise dielectric effects. Oil has T_1, T_2 and proton density (PD) values that are significantly different from that of aqueous solutions and may give rise to a lower MRI signal.

Phantom solutions may also contain other salts to create loading conditions similar to a patient. For more information on suitable filling materials for phantoms, see Section 2.3. This loading increases the noise seen in the detected MR signal (see Section 3.2.4).

3.2.3 RF coil type

The RF coil is a major component in the detection of the MR signal and is therefore very important when considering SNR measurements. Some RF coils are transmit–receive, whereas others are receive-only. Most receive-only coils are multiple-element phased-arrays (simply known as arrays), which utilise a number of smaller coil elements with independent receivers to cover a larger area.

The use of array coils on modern MRI scanners makes SNR a very important measure of image quality, as the loss of signal from one coil element may not be evident from a visual inspection. Some array coils, e.g. for imaging the body, can be mechanically flexible and a consistent physical set-up is crucial to ensure an accurate measurement.

It is important to know whether a coil is transmit–receive or receive-only if a measure of coil loading or coil quality factor, Q, is required. This may be done to compare different RF coils where the same phantom or set-up conditions cannot be used. Q is usually determined from knowledge of either the forward and reflected power through the coil, or the voltage or gain required for a 90° RF pulse (Redpath and Wiggins, 2000), but this can only be done with a transmit–receive coil, where the principle of reciprocity applies (Hoult and Richards, 1976). Corrections for coil loading are not readily applied to SNR measurements with receive-only coils. It should be noted that some modern RF coils may be called transmit–receive, but contain separate transmit and receive coil elements.

3.2.4 Noise from the object

Noise may arise from the RF coil, receiver electronics and the object being imaged. In modern scanners operating at field strengths of 1–3 T with high quality receiver systems, noise is dominated by the object being imaged. Random electromagnetic fields are generated in the object and picked up by the receiver coil (Hoult and Lauterbur, 1979). This is referred to as thermal or Johnson noise, and the standard deviation of these fluctuations σ_{noise} is given by

$$\sigma_{noise} \propto \sqrt{4kT \cdot R \cdot BW} \tag{3.1}$$

where k is Boltzmann's constant, T is the coil temperature, R is an effective resistance and BW is the receiver bandwidth. The effective resistance depends upon the dielectric properties and conductivity of the object, and is linked to how the object loads the RF coil. When making routine SNR measurements, it is important to ensure these properties do not change between subsequent measurements by using a reproducible coil and phantom set-up.

3.2.5 RF system

The RF system is responsible for controlling how much RF energy is delivered to the phantom, as well as collecting, amplifying and digitising the resulting signal. A problem with RF transmission could lead to the wrong RF energy being delivered, which may affect the resulting signal. Issues with the RF receiver could lead to increased noise and/or reduced signal. Modern systems utilise multiple receiver channels and the way these channels are used depends on the active RF coil. A change in SNR could be related to an issue with a specific receiver and not linked to a single RF coil.

3.2.6 Environment

The environment can have an impact on SNR, principally through variations in temperature. Most MRI scanners have air-conditioned rooms which maintain a constant temperature. However, if the bore temperature does change significantly, this can affect SNR through three different mechanisms: changes in the bulk magnetisation, which has a $1/T$ dependence; changes in the noise which, as seen in Section 3.2.4, has a $\sqrt{T}$ dependence; and changes in phantom relaxation times (especially T_1), which have a more complex dependence. Scan room temperatures should be kept constant and phantoms given time to equilibrate if they are not kept inside the room. It may be prudent to record the temperature in the magnet bore to determine whether SNR changes are the result of temperature fluctuations or other factors.

3.2.7 Field homogeneity

RF coils are tuned to deliver and receive RF over a narrow range of frequencies. In the presence of magnetic field inhomogeneities, the Larmor frequency may fall outside this range, resulting in a reduction in the flip angle and decreased signal detection. This will affect the SNR if the inhomogeneity is within the imaging volume.

Most modern scanners will have homogeneities better than 0.5 parts per million (ppm) over a 40 cm diameter spherical volume, such that the effect on SNR will be small. It should be noted that phantoms themselves can introduce small magnetic field inhomogeneities due to susceptibility differences between phantom materials and surrounding air. It is therefore important not to include phantom boundaries when making an SNR measurement.

3.2.8 Reconstruction

The reconstruction process takes raw *k*-space data and converts it into an image. During this process, filters may be applied that reduce noise at the expense of resolution. It is important to understand how MRI scanners reconstruct images and what filters may be applied during this process. Techniques that are used to speed up imaging such as reduced *k*-space sampling, parallel imaging and partial Fourier acquisitions all impact on the noise characteristics and SNR (Dietrich *et al.,* 2008). These and any pre-reconstruction filters should be disabled when making SNR measurements, unless the point of the measurement is to assess the impact of using these techniques. Some MRI scanners apply default filters that are difficult to disable. In these situations, SNR can be calculated with the filter in place, but it should be noted that a filter has been used, especially when comparing different scanner manufacturers and models.

Some scanners apply filters that 'warp' the final image to correct for gradient-induced geometric distortions. These filters are non-linear and will modify both the signal and noise characteristics in the periphery of large FOV images (Figure 3.1). This filter should be disabled, particularly when making measurements of noise from the background with large FOV images.

3.3 Measurement of SNR

3.3.1 Review of methods

Given that both the signal and the noise can be affected by so many parameters, the measurement of SNR has to be performed carefully to ensure no errors are introduced. The most accurate method of measuring SNR is to perform a large number of identical acquisitions and then determine the SNR on a pixel-by-pixel basis by

$$\mathrm{SNR}(x,y) = \frac{\mu(x,y)}{\sigma(x,y)} \tag{3.2}$$

where x and y represent the pixel positions in the image, and $\mu(x,y)$ and $\sigma(x,y)$ are the mean and standard deviation of the signal across all acquisitions. This technique is very

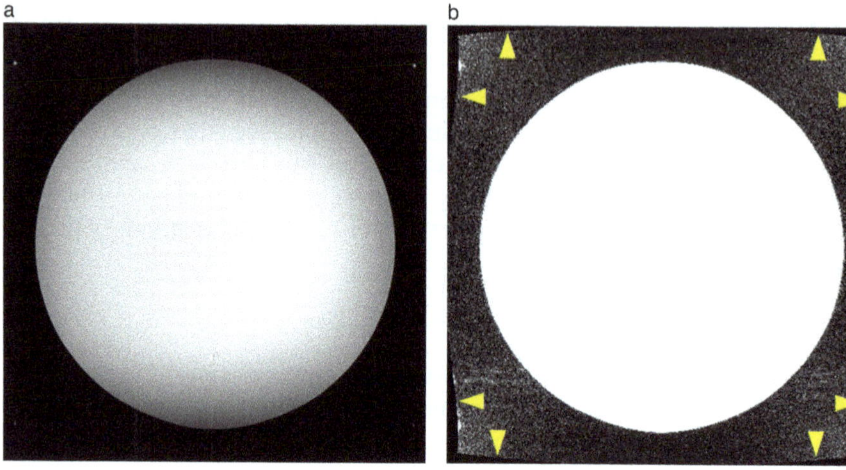

Figure 3.1 (a) A large FOV (450 × 450 mm) image of a flood phantom. (b) Windowed version of the same image clearly demonstrates the 'warping' at the periphery of the image (yellow arrowheads)

time consuming and is not recommended unless assessing SNR in single-shot sequences such as echo planar imaging (EPI).

A quicker method is to only acquire two images and calculate the noise from an image generated by subtracting the two. This is the preferred method in AAPM Report 100 (AAPM, 2010). Dietrich *et al.* (2007) reported that this method gives results that are in agreement with the multiple acquisition method described above. It is essential that both images are acquired under identical conditions. Transmitter and receiver adjustments, shim settings, reconstruction and final pixel scaling factors should be constant for both images. A region of interest (ROI) placed over the phantom position in one of the original images (Figure 3.2a) will give a measure of the signal, with the same ROI placed in the subtracted image (Figure 3.2b) giving the noise. Smaller multiple ROIs may be used if there is evidence of low spatial frequency variations across the subtracted image. SNR is calculated from each ROI and then averaged. However, consistent positioning of individual ROIs may be more challenging than a single larger ROI, particularly if this is being done manually. The noise in the subtracted image ROI will follow a Gaussian distribution (Figure 3.2c) with a mean of zero and a variance equal to twice that seen in the original images.

SNR is calculated by

$$\mathrm{SNR}_{\mathrm{Subtraction}} = \frac{\sqrt{2}\mu_{\mathrm{phantom}}}{\sigma_{\mathrm{subtracted\ image}}} \tag{3.3}$$

where μ_{phantom} is the mean pixel intensity in one of the phantom images, and $\sigma_{\mathrm{subtracted\ image}}$ is the standard deviation in the subtracted image. The NEMA Standards Publication on SNR calculation, MS 1-2008 (NEMA, 2008a), contains an

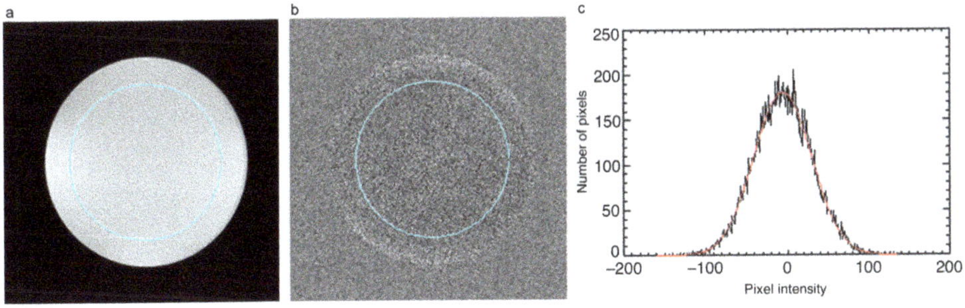

Figure 3.2 (a) ROI placement for a circular phantom in an image. (b) Same ROI position in the subtracted image. (c) Histogram of noise in the ROI demonstrating a Gaussian distribution

alternative method for calculating the standard deviation if temporal instabilities are encountered during acquisition. This looks at the difference between adjacent pixels in the subtracted images to limit the effects of low spatial-frequency variations on the standard deviation. Further details on the temporal instability correction method can be found in MS 1-2008 or Steckner (2010).

A third method only requires the acquisition of a single image, with a measure of the noise taken from the image background. The noise distribution in the image background will not be Gaussian (due to the magnitude reconstruction of MR images) and may, for phased array coils, vary across the image. However, since this measurement only requires the acquisition of one image, it can be used as a consistency check for daily QA, and has been found to correlate with the subtraction method of SNR (Firbank *et al.*, 1999; McCann *et al.*, 2013). In this case, the signal ROI (or ROIs) will be similar to that used for the subtraction method; the noise should be sampled from several regions in the background, taking care to avoid regions that may contain artefacts or ghosting (Figure 3.3a). The noise ROIs are shown here as circles, but could be any shape, as long as they encompass a sufficient area to obtain a good estimation of the noise distribution (Figure 3.3b), and avoid regions of image artefacts (e.g. ghosting and edge ringing).

SNR is calculated by

$$\mathrm{SNR}_{\mathrm{Background}} = \frac{f \times \mu_{\mathrm{phantom}}}{\sigma_{\mathrm{background}}} \tag{3.4}$$

where μ_{phantom} is the mean pixel intensity in the phantom image, $\sigma_{\mathrm{background}}$ is the mean standard deviation from the four ROIs in the image background and f is a factor that corrects for the non-Gaussian nature of the noise distribution. In the case of a single element coil, where the background noise will follow a Rayleigh distribution, this correction factor is 0.66 (Kaufman *et al., 1989*). This assumes that no filters or other processing is performed during image reconstruction.

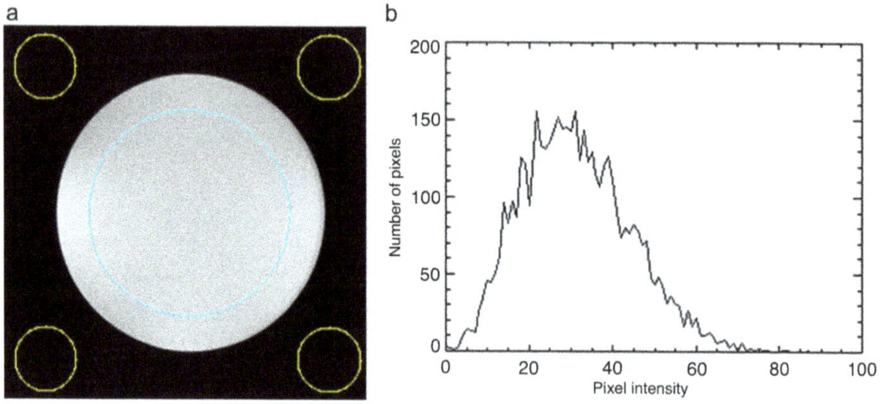

Figure 3.3 (a) Suggested ROI placement for a single image acquisition measurement of SNR. (b) Histogram of the background noise, demonstrating a non-Gaussian distribution

This correction factor is not applicable for multi-element array coils, or situations where raw data filters, signal intensity normalisation, parallel imaging or geometric distortion correction have been used. Constantinides *et al.* (2004) derived correction factors for sum-of-squares array coil reconstruction. However, some scanners utilise an adaptive (spatial-matched filter) phased-array reconstruction algorithm that can significantly reduce background noise levels (Walsh *et al.*, 2000). Correction factors for this reconstruction method have not been explicitly calculated; however, Dietrich *et al.* (2008) describe the shape of noise distributions using different reconstruction methods, and it should be possible to derive correction factors using this information.

In a routine QA programme where the same coil, phantom, imaging sequence and parameters are used, a correction factor is not necessary, unless comparisons are being made with different coils (e.g. comparing an 8-channel body coil with a 32-channel body coil), or with SNR measurements made using the subtraction method. When assessing SNR on specific sequences or clinical images, the background method may be the only practical method of acquisition and it is important to use the appropriate correction factor that takes into account all influences of imaging and reconstruction parameters on the noise distribution.

It is possible to calculate an absolute measure of SNR (ASNR) that removes the influence of sequence parameters and phantom properties (McRobbie, 1996). This is given by

$$\mathrm{ASNR} = \mathrm{SNR} \times \frac{F \times \sqrt{\text{pixel bandwidth}}}{\Delta x \times \Delta y \times \Delta z \times \sqrt{(N_{\mathrm{av}} \times N_{\mathrm{PE}})}} \tag{3.5}$$

where SNR is the measured SNR (ideally using the subtraction method), F is a factor that depends upon the object relaxation times and sequence parameters, Δx and Δy are the horizontal and vertical pixel sizes, respectively, Δz is the measured slice thickness, N_{av} is the number of averages and N_{PE} is the number of phase-encoding steps. For volume RF coils such as the quadrature whole-body coil, this value can be compared to a theoretical SNR that takes into account coil loading, phantom geometry and conductivity. Formulae for this calculation can be found in McRobbie (1996) and Ocali and Atalar (1998).

Other methods of SNR calculation have been documented, including segmenting raw data to synthesise two images from a single acquisition (Steckner *et al.*, 2009), switching off RF transmission to generate an image with no signal components, or applying filters to single image acquisitions to measure the high-frequency noise components (McCann *et al.*, 2013). These methods are not discussed in detail here.

3.3.2 Purpose and frequency of measurement

It was stated earlier that SNR is a useful yet non-specific measure of image quality. This makes it an ideal quantity to use for a frequent QA programme, for example, on a daily basis. However, given the number of different RF coils that come with a typical scanner, it would not be possible to assess SNR for every coil on a daily basis. It is therefore recommended that the coils used most frequently (e.g. head, spine, knee and body/torso/pelvis coils) are tested on a regular basis, with other more specialised coils tested infrequently. A proposed programme for SNR testing of coils is described in Table 3.1. This programme is designed to test the system and coils on a regular basis, without placing an excessive burden on scanner and staff time. This is not meant to be a definitive programme and can be modified to suit the circumstances of each centre. However, it is recommended that some QA is carried out on a daily basis.

3.3.3 Phantoms

The ideal test object for performing SNR measurements is a flood or uniform phantom. As RF coils come in a range of sizes and configurations, it is not possible to specify a single phantom for all measurements. However, general specifications can be made for the testing of the integrated whole body T/R coil, and head coil. NEMA Standard MS 1-2008 suggests that the dimensions of the phantom area seen on an image should be a minimum of 10 cm for a head coil and 20 cm for a body coil (NEMA, 2008a). Some flood phantoms for the whole body T/R coil may require a loading annulus.

Some composite phantoms (e.g. the ACR test object) contain flood field sections, which can be used for SNR measurements.

Manufacturers normally provide flood phantoms for all of their coils, and it is recommended that these are used for routine testing of all phased-array coils. These may be non-standard shapes and sizes specific to the coil, or utilise multiple smaller

Table 3.1 Proposed programme for SNR testing of RF coils as part of a quality assurance programme

Type of test	Details
Acceptance testing, coil replacement and major upgrades	Integrated whole body coil: Perform SNR measurements using the subtraction method in coronal, axial and sagittal planes. Use either the manufacturer's phantom or other suitable phantom. Acquire two images per plane and calculate SNR using the subtraction method. Calculate absolute SNR and compare to theoretical values. For all other coils: Use either the manufacturer's phantom or other suitable phantom. It is recommended for simplicity that the manufacturer's set-up is used. In some cases, the user may decide an alternative set-up is more suitable. In this case, the set-up procedure should be clearly documented. Acquire two images in a single plane and calculate SNR using the subtraction method. Background SNR values can be calculated from these images and used as a reference for daily QA.
Annual QA	Test the integrated whole body coil and all other coils, especially those that are not part of the daily QA programme. Use the same standard set-up that was used for acceptance, acquire two images and use the subtraction SNR image analysis method. Compare with results obtained at acceptance.
Daily/Weekly QA	Focus on the coils most commonly used and cycle through these, testing one coil daily. For coils with multiple sections (e.g. spine coils that cover cervical, thoracic and lumbar regions), it may be advisable to test each region individually over several days. Use the same standard set-up that was used for acceptance and a single image acquisition and background SNR analysis method. Compare with background SNR reference values.

phantoms. Refer to the manufacturer's literature on how to set up and use these phantoms. If different phantoms are being used, the set-up should be clearly documented to ensure consistency.

3.3.4 Procedure

The phantom and coil should be set up in a consistent manner. They should be positioned at the isocentre of the magnet. For certain coils (e.g. knee, shoulder and wrist) that are normally used off-centre, an offset is allowed, on the condition that this set-up can be accurately reproduced. It may be necessary to wait several minutes before commencing the QA measurement to allow the fluid inside the phantom to settle; otherwise the resulting images may contain motion artefacts. Routine positioning scans should be run to ensure the phantom is positioned correctly. The SNR QA measurement should be a single echo Spin-Echo (SE) using the parameters specified in Table 3.2. The parameter names and units are taken from the NEMA DICOM standard part 3 MR image module (NEMA, 2016). Scanner manufacturers may use different terminology or units for some of these parameters. The user should be familiar with their scanner and able to set up a suitable sequence to produce images with these DICOM parameters.

This standardised sequence will take about 4 minutes and 20 seconds to run. For the purposes of daily QA where scanner time might be at a premium, the user can implement an alternative sequence (e.g. fast spin-echo, spoiled gradient-echo, steady-state free precession) with parameters to reduce scan time. The sequence and parameters should be documented and it should be possible to link SNR values back to the spin-echo measurements performed at acceptance.

For array coils, it is convenient to test several elements simultaneously. The orientation of the imaging slice should be selected to visualise output from as many of these elements as possible. For example, use a transverse orientation for an eight-channel brain coil and a sagittal orientation for an eight-channel foot/ankle coil. For some coils (e.g. a body array, or a dedicated breast coil), it may be sensible to have several phantoms within the FOV.

3.3.5 Analysis

Analysis can be performed using any piece of software that is able to draw regions of interest (ROIs) and report back the mean and standard deviation pixel values within the image. Manufacturers will usually provide this functionality on the MRI system, but third-party and in-house software can also be used if it has been validated to ensure accurate results. Positioning of ROIs should be consistent between subsequent measurements as this can introduce additional variability into the results, especially when testing surface coils that have significant non-uniformity. It may be advisable to have written, illustrated protocols to ensure the analysis is done in a consistent manner. If it is not possible to have one large ROI positioned within the phantom, then several smaller ROIs can be used.

Table 3.2 Recommended sequence parameters to use for SNR assessment

Parameter name	Recommended value	Comment
Scanning sequence	Spin-Echo (SE)	
Repetition time (TR)	1000 ms	Depending upon the test material a shorter TR, e.g. 500 ms may be acceptable.
Echo Time (TE)	15–30 ms	A shorter TE, e.g. 15 ms may be more relevant to clinical protocols.
Slice thickness	5 mm	In order to maintain good slice integrity, thinner slices should not be used.
Pixel bandwidth	130 Hz $pixel^{-1}$	If other values are used, the results should be normalised in accordance with Equation 3.1.
Acquisition matrix	256 × 256	The displayed image matrix should also be 256 × 256.
Field of View (FOV)	Variable	FOV can be adjusted to fit the phantom with additional space for background ROIs.
Slice orientation	Variable	Transverse (TR), sagittal (SA), coronal (CO) as required.
Number of averages	1	
Filters	None	It may not be possible to disable some filters.
Parallel imaging	Off	
Percent sampling	100%	
Other options	Off	

3.3.6 Pitfalls

If the measured SNR value lies outside the expected range, it should not automatically be assumed that there is a problem with the system. It is advisable to review the images, check the following and repeat the measurement if necessary.

- Phantom set-up – Has the phantom been set up and positioned correctly? Has enough time been allowed to let the phantom fluid settle?

- Coil set-up – Has the RF coil been set up and connected properly?
- Scan set-up – Has the scan been set up with the correct parameters, slice position/orientation and coil selection?
- Acquisition – Were shim, transmitter and receiver adjustments performed properly? If multiple measurements are being performed, were the adjustments constant between acquisitions?
- Image review – Do the images appear similar to previous QA images? Is there any artefact present that could explain the drop in SNR?
- Analysis – Has the analysis been performed correctly with ROIs in the correct positions?

3.3.7 Interpretation

SNR values should be compared to results obtained during acceptance. It is difficult to set a reference range without performing multiple measurements. Firbank *et al.* (2000) suggested 5 to 10 consecutive scans could be used to calculate a reference SNR value and set the range to the mean $\pm$ two standard deviations. Other unpublished sources have suggested that over 30 measurements are needed to gain an accurate measure of the standard deviation. This approach is not practicable for every coil and is not recommended unless there is a specific need to do so (e.g. QA for a quantitative MRI study extending over several years). However, it should be possible to estimate an acceptable range even with a small number of values and use this as a reference range. Note that SNR may increase as well as decrease.

Further investigations should attempt to isolate what component in the system has caused the change in SNR. This may involve testing other RF coils, different transmit or receiver channels, testing individual coil elements on phased array coils, or using other QA tests (e.g. slice thickness, uniformity, geometry).

3.4 Uniformity

Uniformity is a measure of the ability of the MRI system to produce a constant signal response over the imaging volume. Ideally, the signal received from an object should not depend on its position within the scanner. In practice, this is not the case. RF coils used for transmit require high uniformity to ensure that the flip angle over the imaging region does not vary significantly, as this could adversely affect the signal and contrast. Phased-array receive only RF coils are intrinsically non-uniform and images produced using these usually require normalisation filters to reduce the effects of non-uniformity (Figure 3.4). However, in images with high SNR, the effects of poor uniformity are not usually significant, unless quantitative measurements are being made.

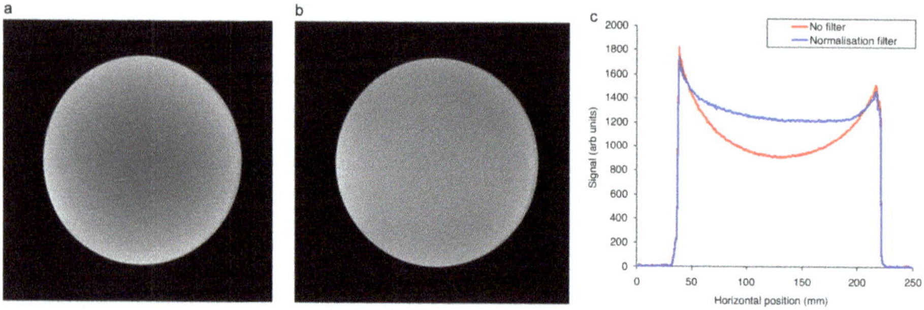

Figure 3.4 (a) Image of an oil-filled phantom scanned using a phased-array head coil. (b) Same phantom but with a normalisation filter applied. (c) Horizontal profiles across the images, showing the effect of the normalisation filter

This section describes the factors affecting uniformity in MRI, how this parameter can be quantified, and the action to take if this parameter is outside specification.

3.4.1 Main field homogeneity

The factors that link magnetic field homogeneity and image uniformity are identical to those for SNR. These are discussed in detail in Section 3.2.7.

3.4.2 RF homogeneity

Transmit RF coils are designed to produce a uniform RF (B_1) field over the imaging volume. However, the RF field will change rapidly at the periphery of the coil, or close to the conducting elements. It should be noted that there is no linear relationship between the RF homogeneity and the final image uniformity, as the image signal depends on the sequence type and parameters. The sensitivity of the receiver RF coil is also a factor; for a transmit–receive RF coil, this will be similar to the transmit RF homogeneity.

The choice of phantom can affect the RF homogeneity, especially at higher field strengths (3 T and above). Phantoms with high dielectric constants can change the RF wavelength to the point where it is comparable with the phantom dimensions, leading to dielectric resonances and RF penetration effects. These manifest as significant non-uniformities in the resulting image (Figure 3.5a). The use of a phantom filled with a material with a low dielectric constant (such as oil) can minimise these effects (Figure 3.5b). See Section 2.3 for further details.

Some modern high field scanners now have "parallel transmit" technology, whereby the whole body coil has multiple elements, each one transmitting RF with an optimised phase to reduce dielectric effects and therefore non-uniformities. Any uniformity measurement of such a system will need to take this into account.

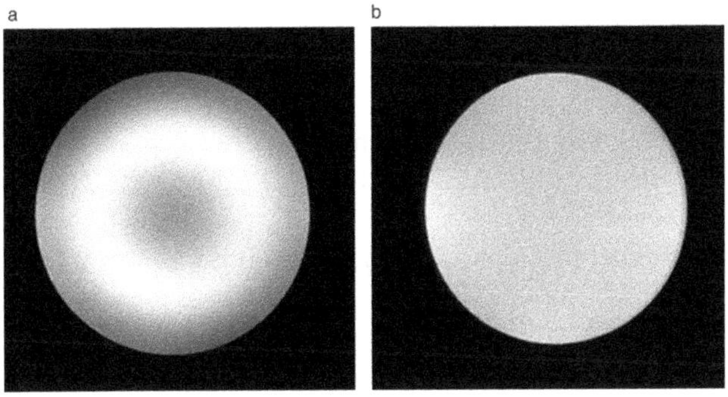

Figure 3.5 Two images of identically sized phantoms taken at 3 T using the integrated whole body coil. (a) Copper sulphate phantom, showing significant non-uniformity due to dielectric effects. (b) Oil phantom with negligible dielectric effects, demonstrating a greater uniformity

It is recommended that array receive-only RF coils such as those now used for spine, body and head imaging are not routinely tested for uniformity unless there is a specific reason (e.g. they are being used for quantitative measurements).

3.4.3 Gradient eddy currents

Induced currents caused by the interaction between ramping magnetic field gradients and the magnet casing or cryostat radiation shields can result in image non-uniformities and distortions. However, this is unlikely to occur in modern scanners with actively shielded gradients.

3.4.4 Pulse sequence dependence

The effect that the pulse sequence has on uniformity is due to the selection of the flip angle and its interaction with other parameters. The flip angle is determined by the B_1 field, but the resulting transverse magnetisation will depend on the sine of the flip angle. With certain sequence types and parameter combinations (e.g. short TRs), the Ernst angle effect (where a flip angle lower than 90° can give a higher signal) can compensate for areas of non-uniformity. The sequence type can also affect uniformity measurements. For example, a Steady-State Free-Precession (SSFP) sequence can produce images with bands of high and low signal intensity. This is due to resonance-offset artefacts and is specific to this sequence. It is therefore very important to use standard sequences and parameters when performing uniformity measurements.

3.5 Measurement of uniformity

3.5.1 Review of methods

The most common method for measuring uniformity is called integral uniformity. This method is recommended by the AAPM (2010), ACR (2005) and NEMA (2008c). This uses a large ROI that encompasses about 80% of the phantom area, avoiding the edges. The ROI should be similar in shape to the phantom. It is known that noise in the image can bias the result to lower values. To reduce the effects of noise, a low-pass filter may be applied to the image before analysis. This may not be necessary if the SNR in the image is sufficiently high. The low-pass filter kernel, which is convolved with the image, should be of the form

$$\text{Low}-\text{pass filter kernel} = \frac{1}{16}\begin{pmatrix} 1 & 2 & 1 \\ 2 & 4 & 2 \\ 1 & 2 & 1 \end{pmatrix} \quad (3.6)$$

Other smoothing filters (e.g. Gaussian filter) may be used if it can be verified that their action is similar to this filter.

Within this large ROI, the maximum and minimum signal intensities (S_{max} and S_{min}, respectively) are identified and the integral uniformity U_{int} is calculated using

$$U_{int} = 100 \times \left(1 - \frac{S_{max} - S_{min}}{S_{max} + S_{min}}\right)\% \quad (3.7)$$

A coil with perfect uniformity would give a U_{int} value of 100%. NEMA MS 3 (NEMA, 2008a) defines a peak deviation non-uniformity measure (N)

$$N = 100 \times \left(\frac{S_{max} - S_{min}}{S_{max} + S_{min}}\right) = 100 - U_{int} \quad (3.8)$$

NEMA document MS 3 (NEMA, 2008a) also suggests the use of greyscale uniformity maps. These provide a visual representation of uniformity, but no quantitative measure. They are most useful where there is significant non-uniformity. A dedicated computer program will probably be needed to create these maps. They are generated by placing an ROI in the centre of the phantom, calculating the mean signal and then assigning pixels a greyscale depending on their value relative to the mean (see Table 3.3).

An example of a greyscale uniformity map is shown in Figure 3.6.

3.5.2 Purpose and frequency of measurement

Uniformity is not a property that needs regular assessment, as changes in uniformity will almost inevitably be associated with changes in SNR. Performing uniformity

Table 3.3 Assignment of pixel values for producing a greyscale uniformity map

Value in original image	Value in greyscale uniformity map
Below –20% of mean pixel value	Black
Between –10% and –20% of mean pixel value	25% grey
Within ±10% of mean pixel value	50% grey
Between +10% and +20% of mean pixel value	75% grey
Above +20% of mean pixel value	White

measurements on array coils that are intrinsically non-uniform will not be very useful unless this information is required for correcting quantitative measurements. It is recommended that all coils used for RF transmission are assessed for uniformity at acceptance, after major upgrades (hardware or software) or RF coil replacement, and on an annual basis.

3.5.3 Phantoms

Flood phantoms are the best test objects for performing uniformity measurements. Ideally, the phantom should fill the coil as much as possible, although in practice, phantom size will be limited by weight. As with SNR measurements, NEMA Standard MS 3-2008 suggests that the dimensions of the phantom area seen on an image should be a minimum of 10 cm for a head coil and 20 cm for a body coil. Flood phantoms for the whole body T/R coil may require a loading annulus. While a cylindrical phantom is ideal for these measurements, a spherical phantom may be used if its diameter is sufficient. See the discussion in Section 2.3 for more information on flood phantoms.

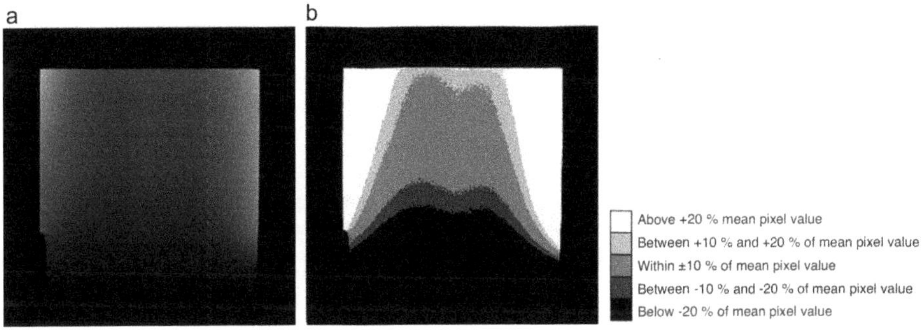

Figure 3.6 (a) Sagittal phantom image taken using an 8-channel head coil. (b) Greyscale map with scale

3.5.4 Procedure

The recommended set-up and scanning procedure for uniformity measurements is identical to that used for SNR (see Section 3.3.4). In fact, images acquired for the purposes of SNR measurements can also be used for uniformity. For the whole body T/R coil and other transmit coils, uniformity should be assessed on transverse, sagittal and coronal images.

3.5.5 Analysis

Integral uniformity may be calculated using the following steps: apply an appropriate smoothing filter (see Section 3.5.1) and place a phantom-shaped ROI that encompasses about 80% of the phantom area. For a circular phantom, this would give an ROI with a diameter of about 90% of the phantom diameter. Find the maximum (S_{max}) and minimum (S_{min}) pixel values within the ROI and calculate U_{int} using Equation 3.7.

In cases where the ROI tool only reports mean pixel value, the following method, suggested by the ACR, can be used. Position a large ROI over the phantom. Manipulate the image so that it only shows black and white pixels. This can be achieved by manipulating the window/level (contrast/brightness) settings. Then adjust the threshold of the image to the point where only a few white pixels are still seen in the ROI (Figure 3.7b). Place a small ROI (diameter of about 5 pixels) over the area of the remaining pixels and note the mean (this will be S_{max}). Adjust the threshold to the point where only a few black pixels are seen in the ROI (Figure 3.7c) and place an ROI in this area to obtain a measure of S_{min}. Use Equation 3.7 to calculate U_{int}.

3.5.6 Pitfalls

Many of the pitfalls associated with SNR measurements (e.g. correct phantom positioning, artefacts, etc.) also apply to uniformity measurements. See Section 3.3.6 for more information about these pitfalls and how to minimise their effect.

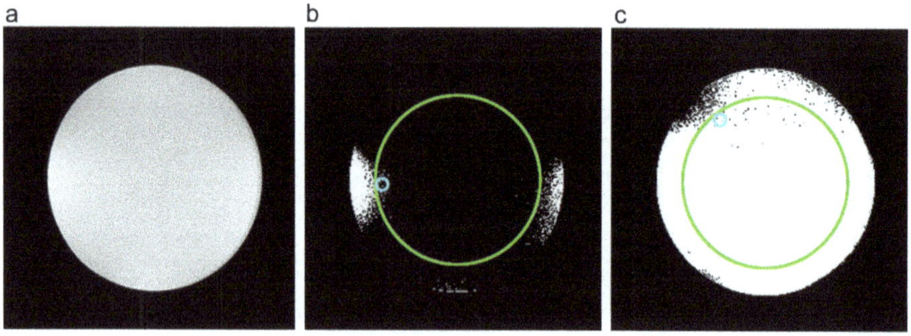

Figure 3.7 (a) Image of a phantom for uniformity assessment. (b) Image showing only black and white pixels, adjusted so that only a few white pixels are seen in the large ROI. (c) Image adjusted so that only a few black pixels are seen in the large ROI

3.5.7 Interpretation

Typically, a whole-body quadrature birdcage RF coil on a 1.5 T scanner would give an integral uniformity of between 80% and 90%. Uniformity in the sagittal and coronal orientations is generally lower than in the transverse orientation due to coil roll-off. The ACR sets its specification for transmit coils as greater than 87.5% for field strengths up to 3 T, and greater than 82% for 3 T and above. However, the AAPM recommends that all integral uniformity measurements should be above 90% for scanners below 2 T. It is recommended that uniformity measurements below 80% warrant further investigation, especially if they are accompanied by a change in SNR.

3.6 Ghosting

In a perfect MR image, there should be a one-to-one relationship between the object being scanned and the final picture. This is often not the case in practice where the presence of noise and other artefacts degrade this relationship. One of the most common artefacts that can affect clinical images is "ghosting", where low intensity copies of the object appear displaced within the image. Ghosting can occur due to a variety of causes, some of which arise from the equipment or sequence parameters and others from the object being scanned. These are described in detail below.

3.6.1 Motion

Motion is the most common cause of ghosting in MR images, although this is usually only seen in patients (Figure 3.8a). The motion of the object being imaged leads to spurious phase variations in the *k*-space data, resulting in ghosts in the final image. The ghosts are translated in the phase-encoding direction and are usually due to bulk patient movement, cardiac respiratory motion or blood/cerebral fluid flow. Other sources include vibrations

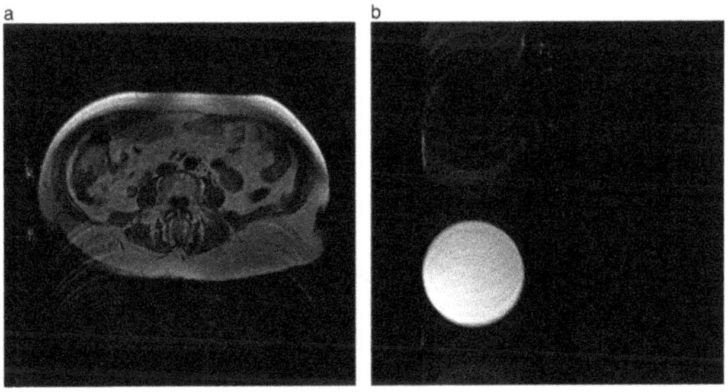

Figure 3.8 (a) Axial image of the abdomen demonstrating ghosting due to motion. (b) Image of a phantom for ghosting assessment, demonstrating similar artefacts. In both cases, the phase-encoding direction is vertical

arising from the equipment or even from external sources. Motion artefacts in phantoms may also be caused by fluid movement (see Figure 3.8b). These artefacts can be minimised by setting up the phantom and then waiting a short time before starting the measurement.

3.6.2 RF and gradient stability

Phase variations in the MRI signal may also be caused by imperfections in generated RF and gradient waveforms, giving rise to ghosting similar to that seen with motion. Modern MRI systems have very stable RF systems and gradient amplifiers, making artefacts arising from this source very unlikely. However, if other causes of ghosting have been ruled out, RF and gradient hardware problems should be considered.

3.6.3 Receiver quadrature errors

The receiving MRI signal is demodulated with a sine and cosine reference signal to produce real and imaginary parts. In older MRI systems, the reference signals were generated using a 90° phase shifter. If the phase shift was not 90°, the resulting image could contain a ghost that is rotated by 180° with respect to the centre of the image (Figure 3.9a). Modern systems using digital demodulation techniques should not give rise to this type of ghost, as the digital reference signals are always 90° out of phase with each other.

3.6.4 Pulse sequence

Some sequences may be more susceptible to ghosting than others. This is certainly true of Echo Planar Imaging (EPI), where imperfections in the rapid switching of the readout gradients can produce a ghost image shifted by half the FOV in the frequency-encoding direction. Multi-echo spin-echo and fast gradient-echo sequences may produce stimulated echoes that produce a single ghost reflected across the centre

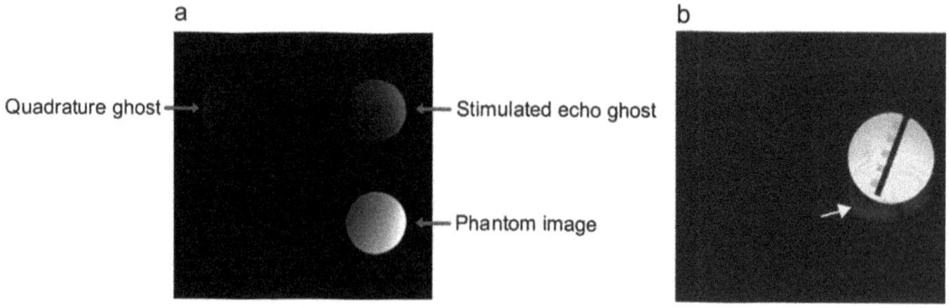

Figure 3.9 (a) Image of a phantom, demonstrating the relative positions of the stimulated echo and quadrature ghost. The phase-encoding direction is vertical. Note this image has been artificially generated to demonstrate the features of each ghosting artefact. (b) Stimulated echo ghost seen in an image of a phantom

of the FOV in the phase-encoding direction (Figure 3.9b). Modern sequence design is tailored to ensure these ghosts are minimised. In the case of EPI, this is normally done using a calibration acquisition; multi-echo spin-echo sequences have symmetric spoiler gradients around the 180° refocusing pulses and fast gradient-echo sequences make use of RF spoiling schemes.

3.6.5 Sequence parameters

Some sequence parameters can affect the degree of ghosting; for example, ghosting artefacts due to movement can be reduced by increasing the number of averages. Some techniques such as parallel imaging may give rise to ghosting under certain conditions. As with other QA measurements, a consistent set-up and imaging protocol should be used.

3.7 Measurement of ghosting

3.7.1 Review of methods

Ghosting is one property that can be easily assessed without the need for measurements of image intensity. A visual inspection of the image should be an adequate way of determining if significant ghosting is present. However, if a quantitative measure of ghosting is required, it can be evaluated relative to the true phantom signal or the background noise level. Lerski *et al.* (1988) recommend an approach using an off-centre phantom, measuring the signal within the phantom (S_p) and the signal in any regions of ghosting (S_g), usually from an ROI placed in line with the phantom in the phase-encoding direction. Figure 3.10 shows the positioning of the ROIs. Ghosting is calculated using the following formula

$$\text{Ghosting} = \frac{S_g}{S_p} \times 100\% \qquad (3.9)$$

Values much lower than 100% would indicate significant ghosting in the image.

A more intuitive measure would be to compare the ghosting level to the noise (S_n), taken using an ROI in line with the phantom but in the frequency-encoding direction (Figure 3.10). Using this method, ghosting can be calculated using

$$\text{Ghosting} = \frac{(S_g - S_n)}{S_p} \times 100\% \qquad (3.10)$$

A non-zero value would indicate when ghosting was significantly greater than noise levels.

The ACR (2005) and AAPM (2010) use the flood field section in their phantom to assess ghosting. However, this method could be done with any flood phantom used for SNR measurements. Five ROIs are used: one in the centre of the phantom to

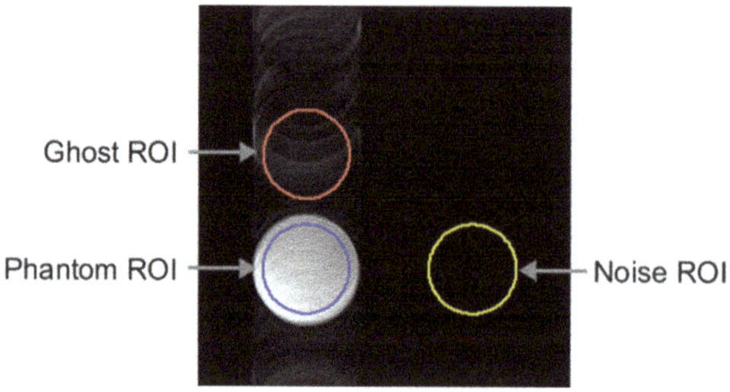

Figure 3.10 Image of an off-centre phantom, demonstrating the position of the phantom (signal) ROI (S_p), the ghost ROI (S_g) and the noise ROI (S_n)

measure the signal, and the other four to the sides of the phantom (Figure 3.11) to measure ghosting and background noise levels. A ghosting ratio is calculated using

$$\text{Ghosting ratio} = \frac{|(\text{top} + \text{bottom}) - (\text{left} + \text{right})|}{2 \times \text{signal}} \tag{3.11}$$

A ghosting ratio close to zero would again indicate ghosting that is not significantly greater than the background noise. It should be noted that this method may not be suitable for detecting quadrature and stimulated echo ghosts.

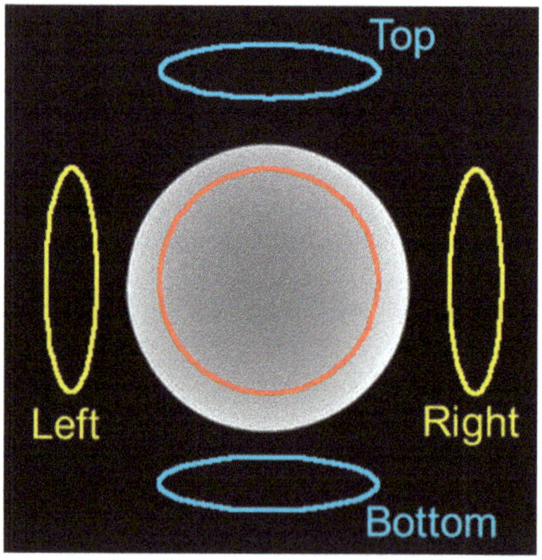

Figure 3.11 Image of a flood field phantom, demonstrating the approximate positions of the signal, left, right, top and bottom ROIs

3.7.2 Purpose and frequency of measurement

A specific measurement of ghosting does not need to be performed on a regular basis, as ghosting can be seen on regular QA images. It is recommended that dedicated ghosting measurements are performed at acceptance, on an annual basis and when unexplained ghosting is present on QA or clinical images. Since ghosting can be sequence-specific, it may be useful to run this test on a variety of sequences, such as EPI, Fast or Turbo Spin-Echo (FSE/TSE), single-shot FSE/TSE (SS-FSE/TSE), Gradient Recalled Echo (GRE) and balanced Steady-State Free-Precession (b-SSFP), particularly if these sequences are being used frequently (e.g. fMRI, diffusion, cardiac MRI, etc.).

3.7.3 Procedure

The best phantom to use for this measurement is a small bottle, offset in the standard head coil. Pads should be used to ensure the phantom cannot move during the measurement. A flood phantom centred in the coil may be used as an alternative. Once the phantom has been positioned at the isocentre of the magnet, remember to allow several minutes for the fluid inside to settle. The standard sequence to test for ghosting is a multi-echo Spin-Echo (SE). The parameters for this sequence are detailed in Table 3.4. If other sequences such as EPI, FSE, SS-FSE, GRE and b-SSFP are being tested, these should use parameters similar to the manufacturer's clinical protocols.

3.7.4 Analysis

The best method for determining whether ghosting is present is to simply review the images, enhancing the contrast in the background if required. Quantitative methods will only be instructive if significant ghosting is seen on visual inspection. If a flood phantom is used for scanning, quantitative analysis should follow the ACR method detailed in Section 3.7.1. The technique using the signal, ghost and noise ROIs (i.e. not the AAPM method) should be used in the case of a small offset bottle phantom.

3.7.5 Pitfalls

The main pitfall when making ghosting measurements is phantom movement, due either to insufficient time between set-up and measurement, or vibrations causing the phantom to move during scanning. Taking care during set-up to ensure the phantom cannot move, and leaving sufficient time (a few minutes) between set-up and scanning will ensure this can be minimised as much as possible.

3.7.6 Interpretation

If significant ghosting is seen on visual inspection, this may not mean there is an issue with the scanner, especially if it is a sequence which is susceptible to ghosting, such as EPI. Measurements made using both the ACR and the small phantom

Table 3.4 Sequence parameters to use for ghosting assessment

Parameter	Recommended values	Comment
Scanning sequence	Multiple-echo SE	
Repetition time (TR)	1000 ms	
Echo Times (TE)	30/60/90/120 ms	Alternatively a double echo with a PD and T_2-weighted appearance, e.g. TE =15, 100 ms.
Slice thickness	5 mm	
Pixel bandwidth	130 Hz $pixel^{-1}$	Not critical to this test unless combined with SNR measurement.
Acquisition matrix	256 × 256	The displayed matrix size should also be 256 × 256.
Field of View (FOV)	250 × 250 mm	
Slice orientation	Axial	
Number of averages	1	
Filters	None	It may not be possible to disable some filters
Parallel imaging	Off	
Percent sampling	100%	
Other options	Off	

method will be acceptable if the ghosting ratio is less than 0.025 or 2.5%. If ghosting above this level is measured, especially on the short TE image, the first course of action is to repeat the measurement. If ghosting is still present, then the source (e.g. phase ghosting, stimulated echo, and quadrature) should be identified from the images. If instability along a particular gradient axis is suspected, repeating the measurement with the phase encoding in a different direction, or with a different orientation (sagittal or coronal) might help isolate the affected gradient.

4 Geometric parameters: spatial resolution and distortion

Daniel Wilson, Gavin Wright, Katherine Lymer and Donald McRobbie

4.1 Introduction

It is widely recognised that MR images are not always an accurate geometric representation of the object under study (Khoo *et al.*, 1997; Breeuwer *et al.*, 2001; Jezzard, 2009). This image distortion can also affect the spatial resolution of the reconstructed image. Such limitations are usually a result of inhomogeneities in the static magnetic field (B_0), gradient miscalibration and non-uniformities, and eddy currents. The inhomogeneities in the static magnetic field are partly due to the technical limitations of designing a homogenous magnetic field over a useful imaging volume and this homogeneity is further degraded when an object or patient is introduced into the magnetic field. Passive and active shimming are both employed to improve the homogeneity but have limitations, particularly in regions where the magnetic susceptibility of the object or patient changes rapidly, e.g. at air–tissue interfaces or around metallic implants. Some sequences, e.g. Echo Planar Imaging (EPI) can be particularly sensitive to magnetic field inhomogeneity distortion.

Linear gradients are required for accurate spatial encoding and distortions occur when the gradients deviate from this ideal. The gradients are usually more linear at the isocentre and there is a decrease in gradient strength towards the periphery of the imaging volume which is observed as a compression of the image. Almost all MR scanner manufacturers now provide some form of distortion correction algorithm to partially compensate for this. The amount of geometric distortion that is acceptable depends on the application for which the MR images are being generated. Some applications, e.g. treatment planning using MR images require the images to have greater geometric accuracy than when MR images are used solely for diagnostic purposes (Cormack, 2008; Gorny *et al.*, 2006).

Methods have been developed (Price *et al.*, 1990; Lerski *et al.*, 1998; ACR, 2005) to assess spatial resolution and geometric distortion to establish the initial performance of an individual scanner, to allow the longitudinal monitoring of scanner performance and, where appropriate, the generation of correction algorithms (Breeuwer *et al.*, 2001; Kristensen *et al.*, 2008). The aim of this chapter is to review the commonly applied methods for the measurement of image spatial resolution and geometric distortion and present procedures suitable for inclusion in routine quality control (QC) protocols.

4.2 Spatial resolution

The ability of any imaging system to distinguish small objects is described by spatial resolution. Given sufficient contrast-to-noise ratio (CNR), the limiting spatial resolution for a conventional spin-echo/gradient-echo sequence in MRI should be determined by the user-selected pixel size. The MR imaging process involves the direct acquisition of spatial frequency information; the higher the spatial frequencies acquired, the smaller the pixel size and the better the spatial resolution.

The spatial resolution of the image can be affected by any factor that distorts the image (magnetic field inhomogeneity, eddy currents, imperfections in the gradient system) and by filtering of the acquired spatial frequency data. This filtering may be due to post-processing filters or may occur during the acquisition process (e.g. T_2 filtering effects in segmented spin-echo sequences).

4.3 Measurement of spatial resolution

4.3.1 Review of methods

High-contrast spatial resolution may be assessed either visually or through quantitative investigation of the modulation transfer function (MTF), line spread function (LSF) and related measures. Visual assessment involves imaging small structures such as line pair patterns or a hole array of a specific size and spacing that are close to the resolvable threshold. A single figure of merit for performance in terms of the limiting high-contrast spatial resolution is produced. In the case of visual assessment, spatial resolution is quantified in terms of the number of line or hole pairs per centimetre of those structures on the limit of spatial resolution.

The MTF describes how the range of spatial frequencies of an object are modulated during the image formation process and provides a more complete description of imaging performance than does a single value of limiting spatial resolution. The MTF can be calculated by taking the Fourier transformation of the line spread function (LSF), which itself may be derived from measurement and differentiation of the edge response function (ERF) (Figure 4.1). The ERF is obtained by taking a profile across a sharp high-contrast step (e.g. the edge of a block). Typically, the pixel matrix is too coarse to allow the ERF itself to be sufficiently sampled, but in a method described by Judy (1976), the frequency of sampling is increased by orientating the edge at a shallow angle relative to the image matrix. Using this method, a number of profiles should be taken across the edge and a single profile reconstructed by combining the data from each and taking account of the angle of the edge. Due to the angling of the block relative to the matrix, the distance to the edge will vary between corresponding pixels from different profiles. For neighbouring profiles, this difference is simply

$$\partial x = \Delta y.\sin\theta \tag{4.1}$$

where Δy is the pixel dimension perpendicular to the profiles (i.e. the distance between neighbouring profiles) and θ is the angle of the block relative to the matrix (Figure 4.2).

A simpler variation of the Judy method may alternatively be adopted to give similar results, but with a smaller number of points sampling the ERF. Rather than plotting multiple profiles along the matrix direction approximately perpendicular to the edge, a single profile crossing the edge can be plotted along the matrix dimension approximately parallel to the edge (Coward *et al.*, 2007). In the direction perpendicular to the edge, the sampling interval of this profile is as Equation 4.1 and directly gives the ERF in this direction.

Whichever method is used to determine the ERF, the LSF should subsequently be determined by differentiation of the ERF and then the MTF determined numerically as the normalised modulus of the Fast Fourier transform of the LSF:

$$MTF(f) = \frac{1}{norm}\left|FFT\left[\frac{d}{dy}ERF(y)\right]\right| \tag{4.2}$$

where f is the spatial frequency given in increments of

$$\Delta f = 1/(N.\Delta y.\sin\theta) \tag{4.3}$$

where the length of the ERF is N pixels.

The maximum spatial frequency is

$$f_{max} = 1/(2.\Delta y.\sin\theta) \tag{4.4}$$

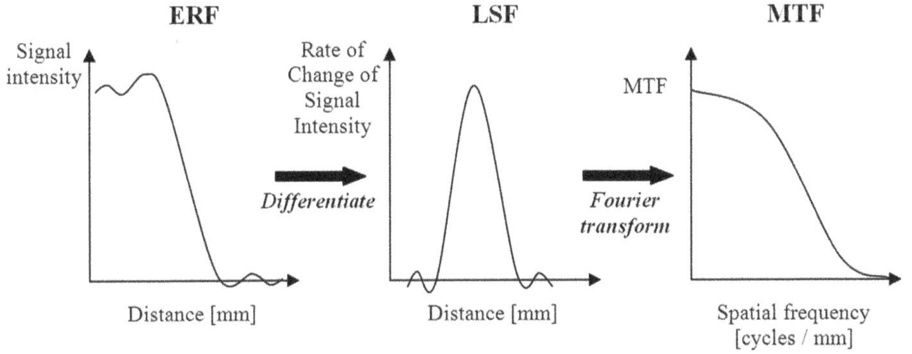

Figure 4.1 The MTF may be derived by first measuring the ERF across a high-contrast edge. Differentiating the ERF gives the LSF which can be Fourier transformed to give the MTF

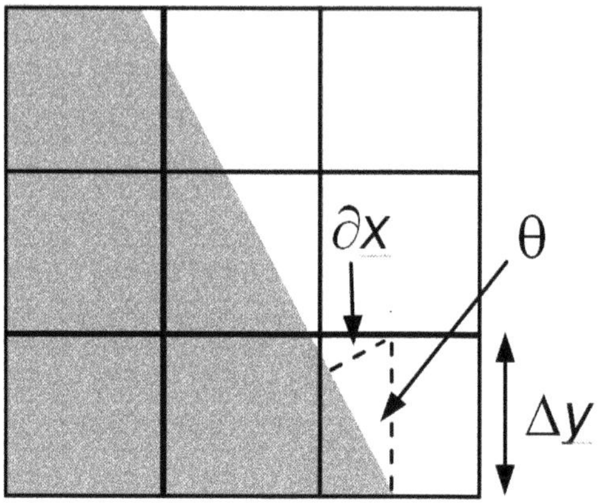

Figure 4.2 Judy method of obtaining the edge response function (ERF)

Use of the MTF requires real-valued, rather than modulus images (see Section 4.3.6) otherwise the asymmetry of the LSF causes distortions in the MTF (McRobbie and Quest, 2002). For examining the spatial resolution for standard clinical sequences, the full width half maximum (FWHM) of the LSF can be used.

An alternative approach is to measure the Contrast Response Function (CRF) from the signal modulation across a set of line pair test patterns (Figure 4.3). The modulation is defined as

$$M_{lp} = a_{lp}/m_{lp} \tag{4.5}$$

where a_{lp} is the amplitude of a line profile across the pattern and m is the mean value of the profile. The CRF is given by M_{lp} plotted against the line pairs. Unlike the MTF, the CRF is not independent of the phantom and can only be considered an approximate metric.

4.3.2 Purpose and frequency of measurement

The purpose of this test is to assess the spatial resolution of the system. This assessment should be made during acceptance testing of the system, and in annual QA checks.

4.3.3 Phantoms

Phantoms suitable for visual assessment of spatial resolution contain patterns or arrays of regularly spaced signal-producing elements (see Figure 4.4 for examples). These elements are typically fluid-filled bars or holes, either rectangular or circular in cross-section, which are separated from each other by non-signal-producing material of a distance equal to the diameter of the hole/bar. A typical phantom will

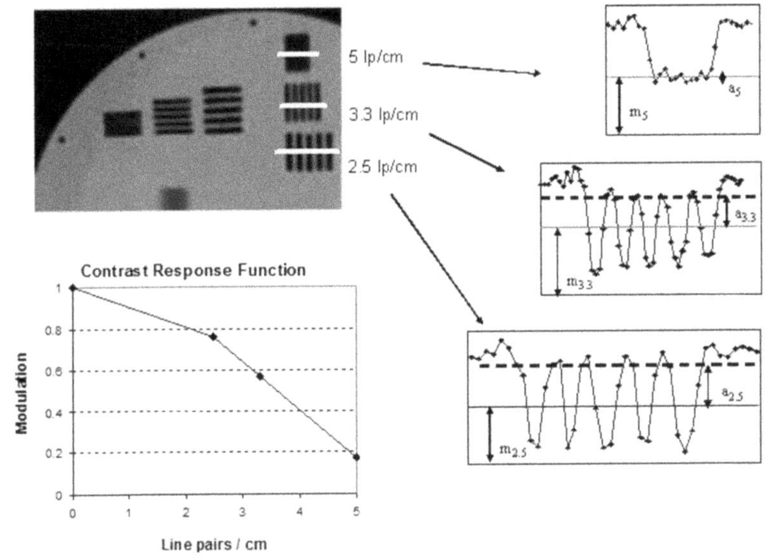

Figure 4.3 Measurements of modulation from line profiles across line-pair test patterns in a phantom and resulting Contrast Response Function (CRF)

contain a number of such patterns in a range of different sizes. The use of bar or line pair details allows for more quantitative analysis of modulation and the CRF.

For the evaluation of the MTF or LSF, a phantom must contain a high contrast edge formed by the boundary between signal-producing and non-signal-producing materials. Typically, this will be a non-signal-producing block inside a fluid-filled

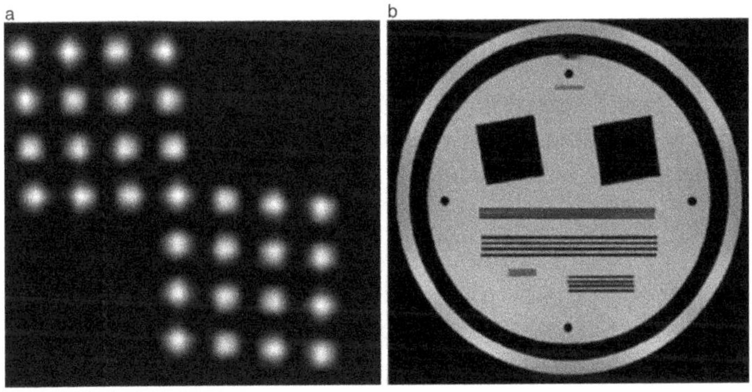

Figure 4.4 Spatial resolution patterns from (a) American College of Radiology (ACR) phantom and (b) Eurospin phantom. Note the MTF blocks displayed in (b)

region, or vice-versa. Such a block will be aligned at a known angle to the pixel matrix to allow evaluation of the ERF by the method of Judy (1976).

4.3.4 Procedure

4.3.4.1 General preparation

- Ensure test objects are in thermal equilibrium with the magnet at room temperature.
- Position the phantom according to the specific user instructions. Unless otherwise directed, ensure the phantom is level and square to the scanner axes.
- A settling time of approximately 5–10 min is recommended before scanning commences to prevent motion artefacts.
- Acquire a localiser or scout image to ensure correct phantom set-up.

4.3.4.2 Test object

Any test object containing spatial resolution bars, holes or grids, or an MTF block.

4.3.4.3 Sequence

A standard spin-echo sequence (not fast spin-echo) is required for visual assessment of test patterns. For MTF measurements, an inversion recovery sequence with real reconstruction should be acquired as this gives a true indication of the signal baseline when measuring the ERF. The exact choice of parameters such as TR and TE are arbitrary but should provide sufficient SNR to analyse the spatial resolution features.

Select a slice thickness that is significantly less than the through-plane dimension of the test pattern (typically 5 mm). Position the slice centrally through the spatial resolution test pattern/MTF block.

Select an acquired pixel size that matches the size of the structures in the test pattern, for example, if the bar phantom contains lines which are 1 mm thick with 1 mm gaps, then a pixel size of 1 mm should be selected.

4.3.4.4 Measurement using test patterns

Align the plane of the test pattern parallel to the scan plane to be investigated with the bar pattern or hole arrays lined up horizontally or vertically.

For a phantom containing bar patterns, two scans should be performed: one with the bars parallel to the frequency-encoding direction, the other with the bars parallel to the phase-encoding direction to allow independent measurements of the spatial resolution in each direction. An array of holes will commonly allow both directions to be measured from a single scan.

4.3.4.5 Measurement of spatial resolution using the MTF method

Align the phantom such that the face of the MTF block lies parallel to the scan plane to be investigated and the block is as close to the quoted angle to the pixel matrix as possible. This angle will usually be specified in the phantom manual. The MTF block will normally allow a measurement of spatial resolution in both frequency- and phase-encoding directions from one scan.

4.3.5 Analysis

A simple method of evaluating the spatial resolution is to perform a visual assessment of a test pattern. The window and level of the displayed image should be adjusted for optimal viewing. The spatial resolution is equal to the size of the smallest test pattern set that can be resolved. At the pixel limit, the limiting resolution can be difficult to judge if there is not perfect alignment of the pixel boundaries with the test pattern. Some phantoms (e.g. the ACR phantom) use multiple rows of holes which are slightly offset to overcome this problem (Figure 4.5).

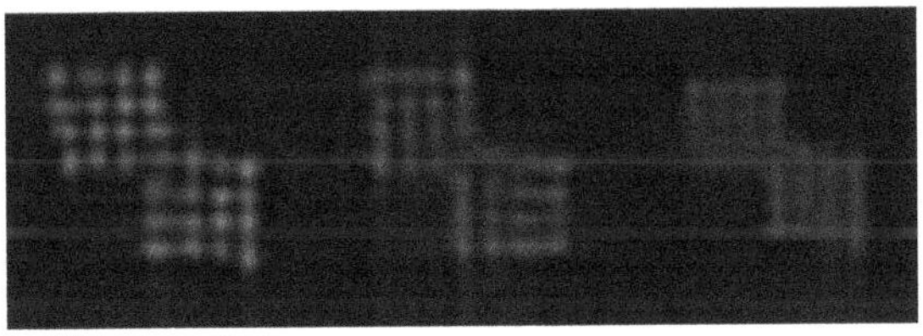

Figure 4.5 Image from ACR phantom taken with a pixel size of 1 mm × 1 mm. The grid of holes in the middle has 1 mm holes and 1 mm gaps and is resolved as is the left-hand grid with 1.1 mm gaps. The right-hand grid of holes (0.9 mm) is not resolved

To help determine whether or not a bar pattern or hole array is resolved, it may be helpful to plot a profile of signal intensity modulation across the pattern or array. Typically, a measured modulation of 50% is considered to be acceptable. If several bar patterns are available, then a CRF can be plotted.

An alternative to visual assessment is to determine the LSF MTF. The MTF can be calculated in the x-direction (or in the y-direction by interchanging x and y) as follows.

1. Select the edge over which to take the profiles. The direction in which spatial resolution is assessed is the one approximately perpendicular to the edge.

2. Measure the angle of the block to the pixel matrix from the image.
3. Calculate the ERF by performing either (a) or (b):
   a) Multiple profiles:
      i. Take several profiles (typically 10) along the matrix dimension in the x-axis.
      ii. Shift the position of each profile by $x = n.\Delta y.\sin\theta$ where n is the number of pixels in the y direction from the first profile and Δy is the pixel size in that direction.
      iii. Combine all shifted profiles into a single profile to obtain the ERF.
   b) Single profile:
      i. Take a single profile along the matrix dimension in the y-axis.
      ii. The effective sampling interval of the ERF is given by $\Delta y.\sin\theta$.
4. Fit the ERF to an appropriate function (e.g. a high order polynomial).
5. Differentiate the resultant function analytically to produce the LSF function.
6. For LSF analysis, calculate the FWHM in pixels and convert to millimetres.

To produce an MTF:

7. Create a set of N evenly spaced values to cover the LSF avoiding edge artefacts.
8. Numerically Fast Fourier Transform the LSF to generate the MTF which should be normalised. The sampling interval of the MTF function will be given by $\Delta f = 1/(N.\Delta y.\sin\theta)$.
9. If the drop-off of the MTF appears undersampled, then regenerate the LSF with more values (e.g. $N = 256$).

A typical measure of limiting spatial resolution is the spatial frequency at 50% amplitude.

4.3.6 Pitfalls

In order to obtain a true measure of limiting spatial resolution, the bars and pixels must be aligned to maximise contrast. By scanning with the bars at a shallow angle to the pixel matrix, the coincidence of the bar spacing relative to the pixel matrix will vary along the length of the bars, and at some position along the bars, the coincidence will be optimal. In this case, it would be reasonable to judge the pattern as resolved even though the bars are not distinguished as separate along their entire length. Scanning with the bars at 45° to the frequency- or phase-encoding directions is possible, but prevents the ability to separately assess spatial resolution in the two dimensions.

When analysing images of an MTF block, unless real (rather than modulus) images acquired from an inversion recovery sequence are used, the LSF obtained will be asymmetric and may not give a true MTF (McRobbie *et al.*, 2007). In this case, a measurement of the Full Width Half Maximum (FWHM) of the LSF can be used as a quantitative indicator of spatial resolution performance.

If the CRF is used rather than an MTF, be aware that it is partially dependent upon the properties of the phantom due to the high spatial frequencies of the bar patterns. For this reason, the CRF underestimates the modulation compared with the MTF.

4.3.7 Interpretation

Spatial resolution equal to the pixel size should be obtainable (AAPM, 2010). Any significant variation from the quoted pixel size (FOV/matrix) should be investigated. This should include looking at all possible causes but especially the inadvertent use of filters, *k*-space undersampling methods or use of segmented pulse sequences.

4.4 Geometric distortion

The term geometric distortion relates to the inability of an imaging system to reproduce the true geometry of a subject. Geometric distortions are an inherent feature of MR imaging systems and, typically, increase with distance from the isocentre. An image may have an error in the scale, i.e. an object is magnified (Figure 4.6b). If any scale errors present are directionally dependent, this will lead to aspect ratio errors that cause the

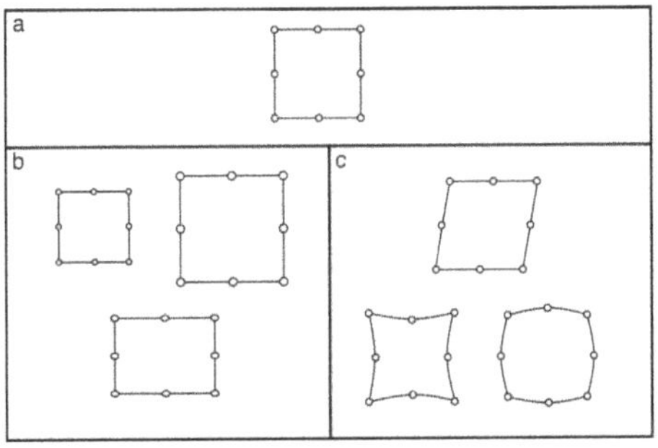

Figure 4.6 Illustrative examples of distortions of a simple geometric shape. (a) Subject's true geometry. (b) Clockwise from top left: reduction, magnification and aspect ratio error. (c) Clockwise from top: shear, barrel and pincushion distortions

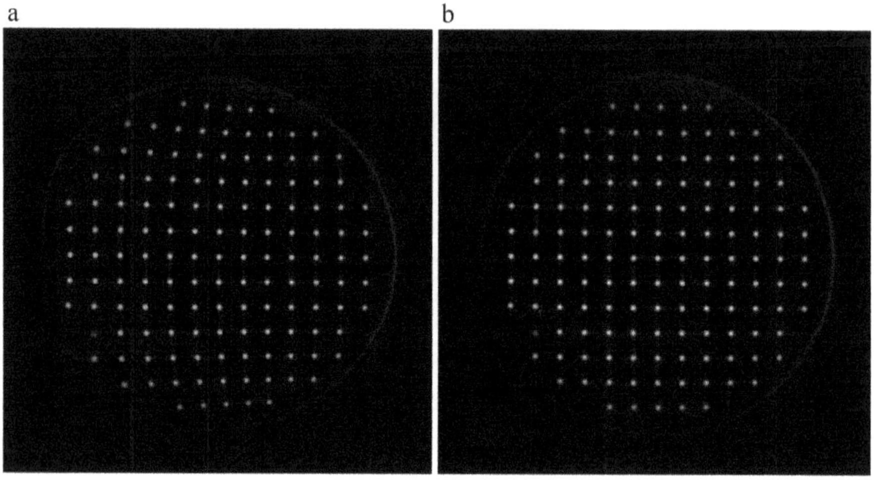

Figure 4.7 Distortion phantom image (a) without distortion correction and (b) corrected image

image to appear stretched (Figure 4.6b). In addition, the magnitude of the error between a point and its true position often depends on the location of the point and this can lead to distortion effects such as pincushion, barrel and shear distortion (Figure 4.6c).

Almost all MR manufacturers provide geometric distortion algorithms which, although unable to correct all distortion, improve the geometric accuracy of the images (Figure 4.7). When performing MRI QC measurements, careful attention is required as to whether geometric distortion correction algorithms are activated. If an assessment of the MRI scanner distortion is required, then it is recommended that images are acquired with and without distortion correction algorithms applied. For routine monitoring of geometric distortion and characterisation of the geometric distortion associated with particular applications, then the correction algorithms should be switched on or match the application under investigation.

4.5 Measurement of geometric distortion

4.5.1 Review of methods

The phantoms used for measuring geometric distortion can be divided into two types: those that contain structures that can be located in two dimensions only and those that contain structures that can be localised in all three dimensions. The 2D structures (e.g. Figures 4.8 and 4.9a) normally form a geometric shape (e.g. a square or a circle) and the 3D structures usually form a set of well defined points (e.g. Figure 4.9b and 4.9c).

a b

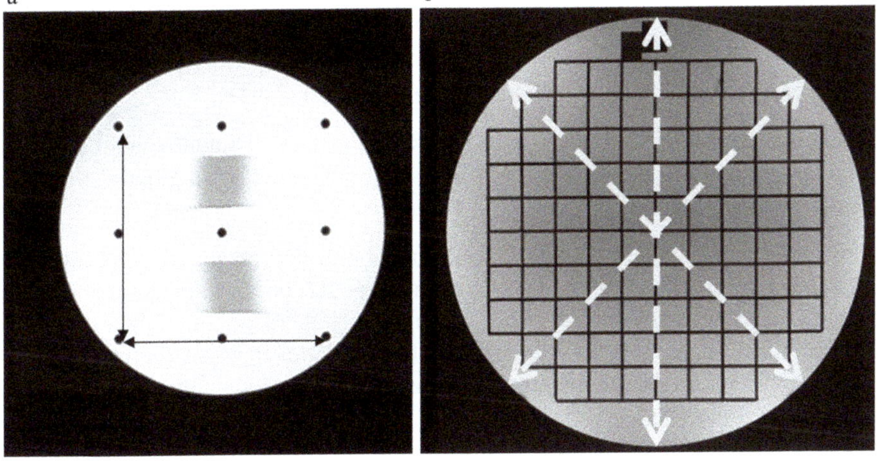

Figure 4.8 Example measurements shown as straight lines on (a) MagNET test object, (b) phantom diameter measurement on ACR phantom (although the grid can also be used)

For 2D structures, the average distance measurement error, maximum distance measurement error and the standard deviation of the percentage distance measurement error are often reported either as a single figure or for both the frequency- and phase-encoding directions (see Figure 4.8 for examples of typical measurements made).

The same method of analysis for structures with defined 3D locations can also be used although the ability to localise points in all three directions also allows an assessment of the positional measurement error between points and their physical location (Wang *et al.*, 2004). The physical location can be obtained from the phantom manual but it is recommended that this is verified by taking a CT scan of the phantom. Some form of

a b c

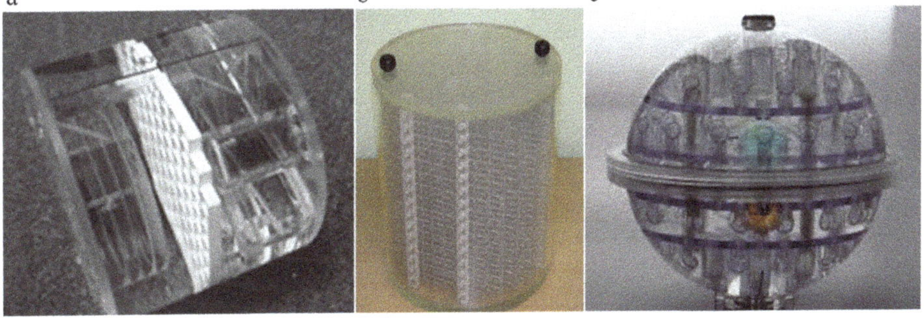

Figure 4.9 Examples of three different phantoms that allow a measurement of geometric distortion. (a) ACR phantom showing the grid, (b) a fully 3D phantom developed by Wang et al. (2004) with 5400 data points and (c) MagPhan® EMR051 (picture courtesy of The Phantom Laboratory, Salem, NY, USA)

rigid body registration between the true locations and the MR locations may be necessary to account for any set-up errors associated with the phantom.

4.5.2 Purpose and frequency of measurement

The purpose of performing QC is to assess geometric distortion against action criteria, to detect changes in the performance of the system and to quantify the error in spatial position of points at different locations.

The frequency and complexity of the measurement will depend on the application for which the MR scanner is being utilised. MR scanners that produce images for diagnosis where geometric accuracy is not critical require only a simple, infrequent check on geometric distortion. In practice, this may be best undertaken using the phantom and images acquired for other QC measures such as SNR, e.g. by measuring the phantom diameter at several angles.

For applications where geometric accuracy is particularly critical, e.g. radiotherapy, radiosurgery or surgical treatment planning, or MR-guided interventional procedures, then it is advisable to have, as a minimum, a simple check on the geometric distortion on an individual patient basis. There are a variety of methods by which this can be achieved such as using fiducials in known geometrical locations round the object or assessing registration quality with a geometrically accurate imaging modality (e.g. CT). An extensive assessment of distortion should be carried out regularly (e.g. annually or when hardware or software modifications are undertaken that could impact on the geometric accuracy) using a distortion phantom that contains 3D structures.

4.5.3 Phantoms

The structures provided by phantoms are usually either the edges of the phantom, plates, rods, grids, spheres or points. Phantom edges, plates, rods and grids only allow an assessment of 2D distortion albeit this can be done in different planes by changing the orientation of the phantom. Phantoms are available that contain 3D structures such as grid intersections (Wang *et al.*, 2004) and spheres. Typical phantoms are shown in Figure 4.9.

Commercially available phantoms are usually designed to fit into the head coil and so only allow a distortion assessment over a small central volume. For applications that require larger field of views, consideration should be given to larger phantoms if available.

4.5.4 Procedure

Set up the test objects as described in Section 4.3.4 and in addition:

- If the structures within the phantom form a straight sided shape, then these sides should be parallel and perpendicular to the frequency-encoding direction.

- The phantom may need to be repositioned several times if it only provides a subset of the data required for an assessment of geometric distortion. This repositioning may involve changing the orientation of the phantom to allow an assessment of distortion in different planes or it may involve rotating the phantom.

4.5.4.1 Sequence

A conventional spin-echo sequence is recommended for routine QC of an MR scanner. The sequence must provide adequate SNR and spatial resolution to robustly identify measurement points. This usually requires a pixel size of 1 mm or less. The slice thickness should be less than half the through-plane dimension of the test pattern and typically not greater than 5 mm. Sequences used for other aspects of scanner QC such as SNR can often be used to assess geometric distortion as well. Routine QC is usually carried out in the head coil and an appropriate FOV to cover the phantom should be selected.

In situations where an application that is dependent on geometric accuracy utilises a specific sequence, then this sequence should undergo regular testing for geometric distortion.

4.5.5 Analysis

All images should be assessed visually and the presence of any gross distortions noted.

For 2D structures, a quantitative assessment should be made based upon the measurement of a number of known distances. These can be performed using either the analysis tools provided on the scanner or off-line image processing software. For guidance on which specific structures should be used for distance measurements, it is recommended that the user refers to the phantom manufacturer's own instructions. Figure 4.8 shows typical measurements for two common phantoms.

Multiple measurements should be made between several pairs of equally spaced structures in two dimensions. Measurements are made in both the frequency-encoding and phase-encoding directions (Figure 4.8a), or in multiple directions (Figure 4.8b) depending on the type of phantom utilised. For any single measurement, d, the distance measurement error (Δ, referred to as linearity in the previous IPEM Report 80) relative to the known separation between structures, l, is given by

$$\Delta = \mathrm{d} - l \tag{4.6}$$

The average distance measurement error, maximum distance measurement error and coefficient of variation of the distance measurements should be reported where the CoV is given by:

$$\mathrm{CoV} = \frac{\sigma_d}{<d>} \times 100\% \tag{4.7}$$

where σ_d is the standard deviation of the measurements and $<d>$ is the mean measured value. If multiple measurements have been made in the frequency- and phase-encoding directions, then results should be reported for each direction separately. For multiple radii or diameter measurements of a circular structure, a single set of results should be reported. In the previous IPEM Report 80, the CoV was referred to as distortion (Lerski *et al.*, 1998).

The measurements described above for 2D phantoms can also be applied to 3D phantoms that contain a number of points. For more complex analyses involving calculating the positional measurement error between an image point and its true location, automated computer analysis is advisable. This may be developed in-house or may be available when the phantom is purchased. As a minimum, the maximum positional measurement error and its location should be reported but useful additional performance statistics include the maximum positional measurement error as a function of distance from the isocentre.

4.5.6 Pitfalls

Most available phantoms fit into the head coil and only assess geometric distortion over a small central field of view. Larger phantoms do exist for evaluating geometric distortions over larger fields of view. They are usually very heavy, cumbersome and expensive.

The 2D phantoms allow a quick assessment of geometric distortion without the need for specialist software. If geometric distortion in orthogonal planes is required, then the phantom usually requires repositioning and the image data acquiring for each plane. This adds significantly to the time required to collect the data.

The described manual analysis procedures may have some intra- and interobserver variability in defining the distance to be measured which may lead to small errors which are not due to any distortion within the image. Automated techniques may improve the measurement repeatability, accuracy and efficiency of the analysis but will not be available to all centres. Where there are sufficient reference points, it may also be possible to extend these automated techniques to provide methods of "un-distorting" the images. However, such approaches usually rely on the expertise of an MR Physicist and the availability of an off-line processing workstation/PC and may not be practical in a clinical setting.

A fundamental limitation of these approaches is that it is difficult to separate the causes of any geometric distortions. Although it is possible to evaluate magnetic field homogeneity separately (e.g. using phase mapping techniques), input from other QC tests and/or the manufacturer's engineer is likely to be required to identify the exact source of persistent geometric distortions.

4.5.7 Interpretation

For a superconducting, closed bore system, the maximum distance measurement error within 9.5 cm of the magnet isocentre should be less than 1 mm (ACR, 2005) or less than 2% over a head FOV (AAPM, 2010). This will increase as the distance from the isocentre increases. It is recommended that the coefficient of variation of the distance measurements should be less than 2% over the same 10 cm distance (Lerski *et al.*, 1998). These figures will usually decrease if a distortion correction algorithm is applied to the acquired images.

Specific applications such as radiosurgery may have their own clinical tolerances which should be adhered to.

Clinically, assuming the distortions in MR images are small and temporally invariant, they can have limited effects – particularly in one-off, routine radiological examinations. However, geometric distortions take on a considerable significance in high-precision applications (such as radiotherapy treatment planning and radiosurgery) and sources of distortion that would have seemed otherwise small (e.g. susceptibility), have increasing relevance. This is also true when combining data from multiple scanners in multicentre imaging studies, where it can be advantageous to evaluate and correct for geometric distortions. Therefore, it is important not only to have a comprehensive QC protocol including regular measurements of geometric distortion, but also to have established action levels based on local clinical applications and past scanner performance.

5 Slice parameters: position, width and profile

Aaron McCann, Anna Barnes and Andrew Patterson

5.1 Introduction

Slice parameters (position, width and profile) are distinct from but related to the signal and geometric parameters described in Chapters 3 and 4, respectively. As with within-slice geometric distortion, accurate positioning and selection of a slice requires a high-quality static magnetic (B_0) field and well-calibrated gradients. The slice profile – like the uniformity measure – is affected by the homogeneity of the time-varying (B_1) field generated by the RF transmission system. As well as maintaining geometric integrity, accurate slice selection is necessary to effect the desired through-plane detection sensitivity for any given scan. Therefore, the measurement of slice parameters is important in quality control (QC) testing to establish intrinsic scanner performance after initial commissioning, following major upgrades, and as part of a battery of regular, standardised QC tests. With the increasing use of MRI in areas where geometric accuracy is critical, such as radiotherapy and surgical planning, and in providing quantitative biomarkers in large clinical trials and research studies, they should be an essential part of routine QC tests performed on the specific pulse sequences used. Slice parameters are well established within the acceptance testing and quality assurance literature, and standardisation guidelines have been published which recommend the measurement of slice width in diagnostic MRI (IEC, 2007; NEMA, 2010a; AAPM, 2010). However, commercial vendors do not, as standard, supply appropriate phantoms for this purpose and the site must therefore acquire an appropriate phantom. Commercially available phantom configurations suitable for measuring slice parameters (Figure 5.1) incorporate inclined ramps and/or wedges, or crossed rods for measuring slice position over larger fields of view (ACR, 2005; Lerski, 2003a; MagNET, 2006). This chapter sets out a framework by which the MR system may be assessed and highlights some of the factors the user must think about when performing such measurements, as well as addressing some of the more common problems that may occur.

5.2 Slice parameters

5.2.1 Slice position

The slice position is defined as the physical distance between the midpoint of a slice and a fixed reference point defined by the scanner co-ordinate system (normally the magnet isocentre). The QC procedure measures the agreement between the nominal slice position (the distance selected during sequence preparation) and the measured slice position as calculated from appropriately acquired images.

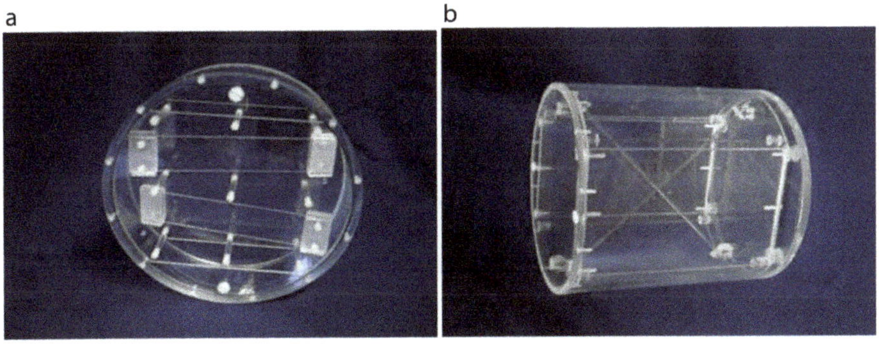

Figure 5.1 Examples of commercial phantoms: (a) inclined plastic ramps for slice width measurement and (b) crossed plastic rods for large field of view slice position measurement (courtesy of David Price, Medical Physics, UCH, London)

5.2.2 Slice profile

The slice profile is defined as the sensitivity of the imaging process along the axis perpendicular to the acquisition plane, and is dependent on the shape and duration of the RF excitation pulse. The desired profile is rectangular, but restrictions on the duration and amplitude of RF pulses mean that ideal profiles are not possible in practice.

5.2.3 Slice width

The slice width (or thickness) is defined for the purposes of this chapter as the full width at half maximum (FWHM) of the slice profile. The full width at tenth of maximum (FWTM) or 'effective slice width' (the integral of the slice profile divided by its value at the slice centre) (Tofts, 2003a) are alternative definitions used. The QC procedure measures the agreement between the nominal slice width (as selected during sequence preparation) and the measured slice width as calculated from appropriately acquired images.

5.3 Factors affecting slice parameters

As with geometric distortion, slice position and width are affected by the quality of the B_0 static field. As inhomogeneity is variable throughout the volume, slice warping (three-dimensional distortion of the acquired slice) will occur to some degree. The use of stronger gradients and higher bandwidth RF pulses reduces the significance of B_0 inhomogeneity effects.

Slice selection accuracy is also dependent on the linearity of the B_0 field when gradients are applied and proper calibration of gradient amplitudes, in concert with the RF transmission pulses' centre ('carrier') frequency and bandwidth. For example, Equation 5.1 describes an axial slice acquired at a slice position z relative to a reference plane (McRobbie *et al.*, 2007):

$$z = \frac{2\pi(f_1 - f_0)}{\gamma G_z} \tag{5.1}$$

where f_1 is the RF carrier frequency (Hz) to acquire a slice at position z and f_0 is the carrier frequency to acquire a slice at reference plane (i.e. $z = 0$). γ is the gyromagnetic ratio and G_z is the amplitude of applied z-gradient (T/m).

The idealised slice width for such an acquisition, Δs, is given by (McRobbie *et al.*, 2007)

$$\Delta s = \frac{2\pi \Delta \upsilon}{\gamma G_z} \tag{5.2}$$

where Δv is the transmit bandwidth of the RF excitation pulse (Hz).

Slice warping (and, to a lesser extent, slice profile) is affected by the homogeneity of the B_1 time-varying field created by the RF transmission system. RF pulses are tailored to best approximate a rectangular slice profile within restrictions imposed by pulse sequence timing or Specific Absorption Rate (SAR) concerns. For example, a specific frequency-modulated pulse scheme first adopts a basic shape (e.g. sinc or Gaussian functions) which is then truncated and apodised (or otherwise filtered) before transmission (Haacke *et al.*, 1999; McRobbie *et al.*, 2007). The design of the pulse is optimised for use with a specific flip angle. The slice profile will often deviate appreciably from a rectangular response when short duration (1–3 ms), selective pulses are required, particularly if they are utilised away from their optimal flip angle (Parker *et al.*, 2001).

Slice profiles are also sensitive to relaxation properties of the phantom material, particularly T_1 for short repetition times (TR) in gradient-echo sequences. For example, when the ratio TR/T_1 becomes less than 1 and the flip angle is large, saturation effects cause significant broadening and other distortions of the slice profile to occur in the regions where the flip angle is varying most, i.e. at the edges of the slice (McRobbie *et al.*, 1986; Tofts, 2003a, p.25). For the purposes of QC, these effects may be minimised using suitably low flip angles, or by ensuring that TR is not substantially shorter than T_1.

For multi-slice 2D imaging, the imperfect nature of slice profiles means that slices adjacent to the selected slice experience partial excitation. If slices are acquired sequentially, then these regions may not have time to relax before their own excitation, causing signal loss at slice edges ('cross-talk'). This effect can be mitigated by introducing a slice gap or using an interleaved slice acquisition procedure.

In addition, any non-coplanarity of slice selection pulses between excitation and readout, or ineffective spoiling of stimulated echoes will also affect the accuracy of slice parameters.

For 3D pulse sequences, through-slice resolution is controlled by applying a second phase-encoding axis in the slice select direction ('partition encoding'), after initial excitation of a thick slab. This approach allows acquisition of much smaller slices as restrictions on RF pulse duration and amplitude do not apply. This has implications for equipment used to measure slice parameters (see Section 5.4.2). McRobbie (1997) has addressed the measurement of "slice resolution" in 3D acquisitions. Through-slice artefacts specific to 3D imaging, such as aliasing and Gibbs ringing, may also compromise measurement procedures.

5.4 Review of methods

The measurement of slice position, width and profile typically requires a fluid-filled phantom containing an angled edge (such as a wedge, a ramp or a rod (Figure 5.2)) made of non-signal producing material (i.e. a 'cold' ramp). Alternatively, an angled, fluid-filled compartment may be embedded in a non-signal-producing block (i.e. a 'hot' ramp). The angled edge cuts through the acquisition plane and projects a profile of the intersection on to one of the in-plane axes. Measurements made from

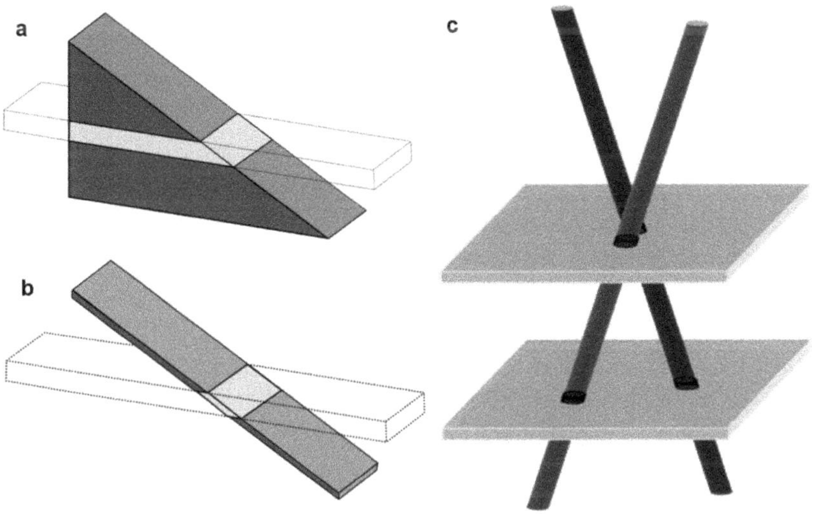

Figure 5.2 Typical configurations of (a) wedge, (b) ramp and (c) crossed rod phantoms with acquisition slices depicted

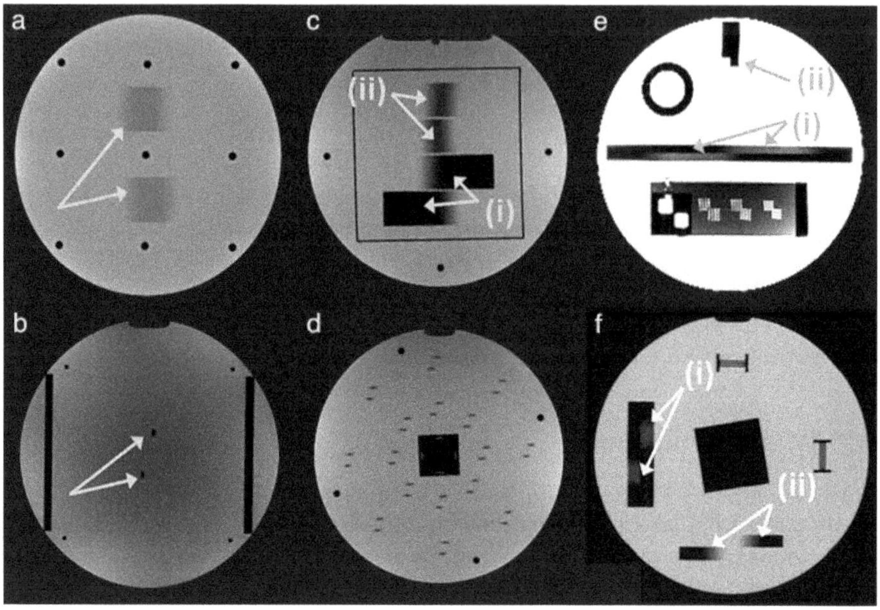

Figure 5.3 Example acquisitions from commercial phantoms with slice width / position test objects indicated. (a) MagNET slice width phantom (cold ramps); (b) MagNET slice position phantom (cold rods); (c) Diagnostic Sonar Ltd Eurospin TO2 phantom ((i) cold wedges and (ii) cold ramps); (d) Diagnostic Sonar Ltd Eurospin TO3 phantom (multiple cold rods); (e) ACR large MRI phantom ((i) hot ramps and (ii) cold slice position wedges); (f) Leeds Test Objects MagIQ phantom ((i) hot ramps and (ii) cold wedges). (c) and (d) courtesy of Claire Davies, Diagnostic Radiology Medical Physics, Poole Hospital NHS Foundation Trust

a single edge are acutely sensitive to errors in positioning the phantom with respect to the scanner's principal axes. Most commercial phantoms contain two such edges inclined at equal and opposite angles to the acquisition plane, which enables compensation for misalignment errors. Figure 5.3 shows acquisitions from various commercial phantoms used for slice selection QC.

Sections 5.4.1 to 5.4.3 describe the procedures involved in determining slice parameters from an axial slice using each of these test object configurations. For more detailed instructions, the NEMA Standards Publication (NEMA, 2010a) is recommended.

5.4.1 Wedge methods

Figure 5.4 illustrates an acquisition through a wedge with inclination angle θ immersed in signal-producing fluid.

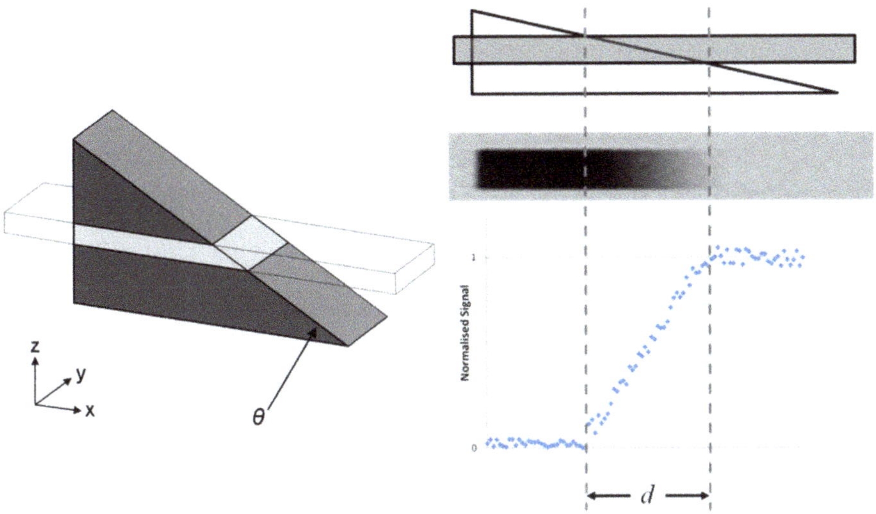

Figure 5.4 Axial acquisition through a cold wedge test object. The edge response function (ERF) of width d corresponds to the intersection of the angled edge with the acquisition plane

5.4.1.1 Measurement of slice width and profile

The edge response function (ERF) of width d projected by the wedge demonstrates a linear increase as the proportion of the acquired slice occupied by the wedge decreases. The measured slice width W is given by

$$W = d \tan \theta \tag{5.3}$$

Differentiating the ERF and rescaling the axis (multiplying by tan θ) gives the slice profile as defined in Section 5.2.2 (Figure 5.5). It may be preferable to estimate the slice width by measuring the FWHM of this profile.

For phantoms with oppositely inclined wedges, two ERF width measurements, d_1 and d_2, may be made from a single acquired image. These may be combined (Equation 5.4) to give an estimate of slice width, W, which compensates for the acute effects of rotational misalignment about the y-axis (AAPM, 2010):

$$W = \frac{(d_1 + d_2)\cos\phi + \sqrt{(d_1 + d_2)^2 \cos^2\phi + 4 d_1 d_2 \sin^2\phi}}{2\sin\phi} \tag{5.4}$$

where ϕ is the angle between the wedges' inclined surfaces (i.e. $\phi = 180–2\theta$).

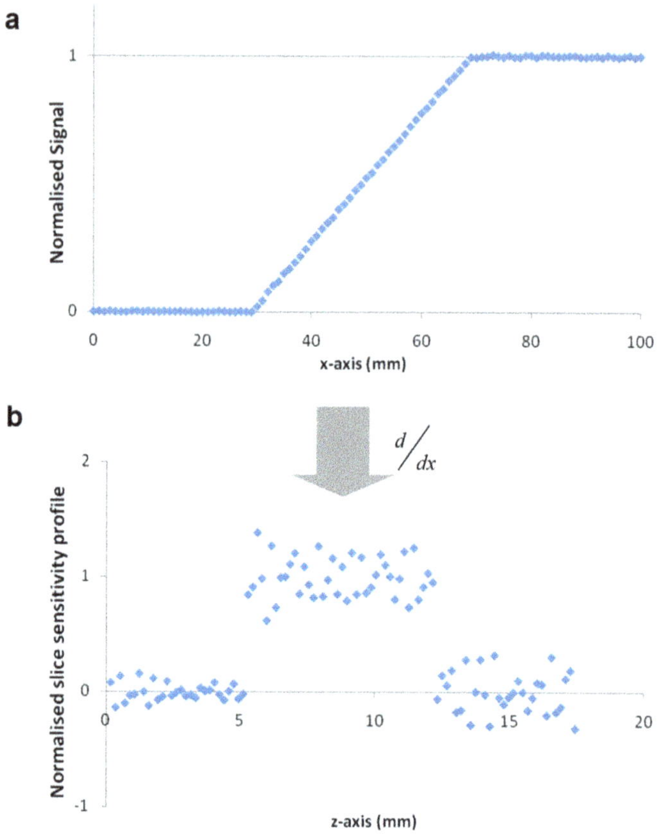

Figure 5.5 Differentiating (a) the ERF of the projected wedge gives (b) the sensitivity profile perpendicular to the acquired slice. Conversion of the axis scale to the through-slice dimension requires multiplication by tan θ (θ = 10° in this example). Note that, in unsigned 'magnitude' acquisitions, noise is rectified in regions of zero signal. Differences in noise distribution are more readily apparent in the differentiated profile

Assuming careful alignment of the phantom, a good approximation of Equation 5.4 may be obtained by separately inserting d_1 and d_2 into Equation 5.3 and taking the geometric mean of the resultant values of W:

$$W \approx \sqrt{W_1 W_2} = \sqrt{d_1 d_2 \tan^2\theta} \quad (5.5)$$

Accurate measurement of slice width may be compromised if there is substantial noise in the image. Noise in the signal profile may be reduced by averaging multiple wedge ERFs from a single image. However, these profiles must be carefully co-registered before averaging to prevent artefactual broadening of the ERF as a result of in-plane rotational misalignment of the phantom.

The slice sensitivity profile may be further assessed in terms of its similarity to an ideal rectangular function. Appropriate metrics (e.g. flatness or sharpness of profile) are not detailed here, but it should be noted that any metrics devised would be acutely sensitive to in-plane spatial resolution settings, SNR, background signal non-uniformity, inaccuracies in phantom positioning and T_1 effects if TR is not sufficiently long.

The wedge method may also be used to generate a 3rd dimension Line Spread Function (LSF; see Chapter 4) for the 'slices' direction in 3D images (McRobbie, 1997).

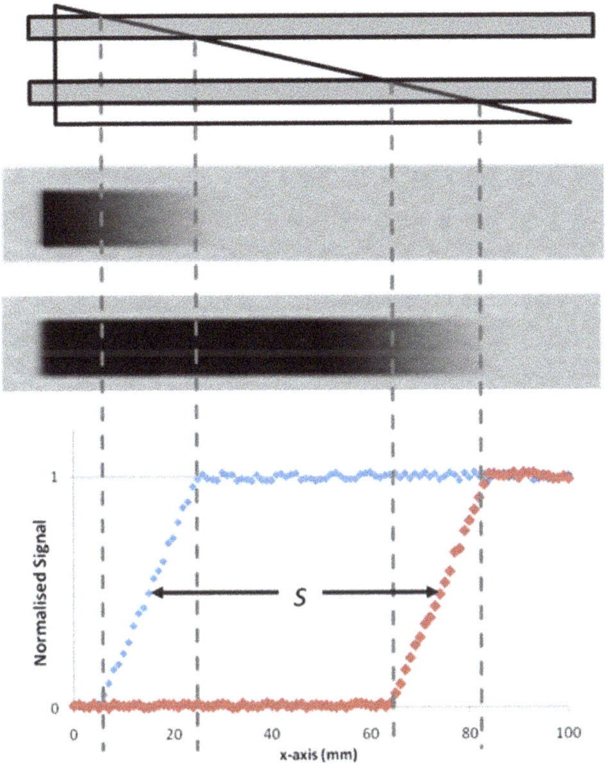

Figure 5.6 Performing two identical acquisitions at separated slice positions results in images whose ERFs are shifted by a distance s relative to each other. The relative slice position, P, is given by P=s tan θ. As with slice width, the presence of opposing wedges allows compensation for rotational misalignment of the phantom

5.4.1.2 Measurement of slice position

Wedges may also be used to assess the accuracy of slice position. Figure 5.6 illustrates how this is enabled by acquisition of multiple slices through a wedge. Measurements may be obtained across a greater field of view (FOV) by acquiring images perpendicular to the long axis of the wedge instead of parallel to it (ACR, 2005).

5.4.2 Ramp methods

Figure 5.7 illustrates an acquisition through a cold ramp with inclination angle θ immersed in signal-producing fluid.

5.4.2.1 Measurement of slice width

The signal profile demonstrates a linear change in regions where the ramp lies partially within the slice, and remains flat where the full ramp lies within the slice. The measured slice width W is again given by:

$$W = d \tan \theta \tag{5.6}$$

In this instance, d is the FWHM of the signal profile. Note that attempts to derive the FWHM from this profile using best-fit curves other than a trapezoidal model may introduce systematic errors. For instance, use of a Gaussian fit is likely to underestimate the true FWHM by approximately 10%.

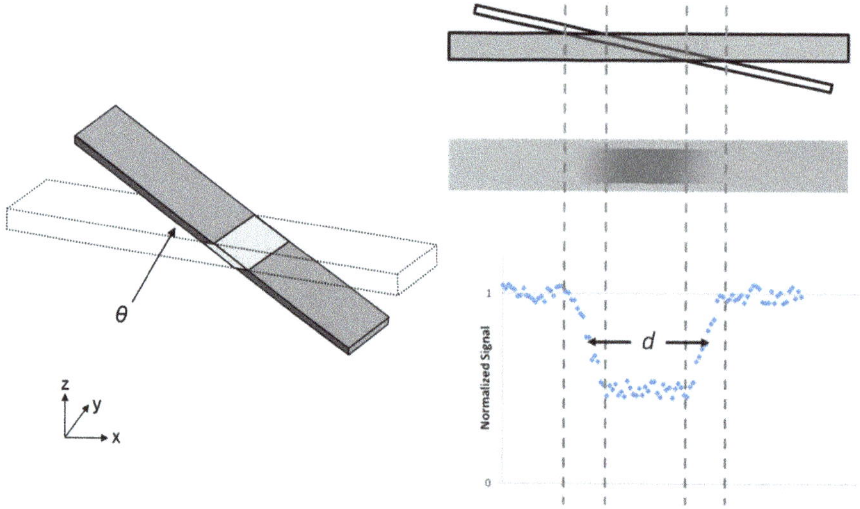

Figure 5.7 Axial acquisition through a cold ramp test object. The trapezoidal signal profile with full width half maximum d corresponds to the intersection of the ramp with the acquisition plane

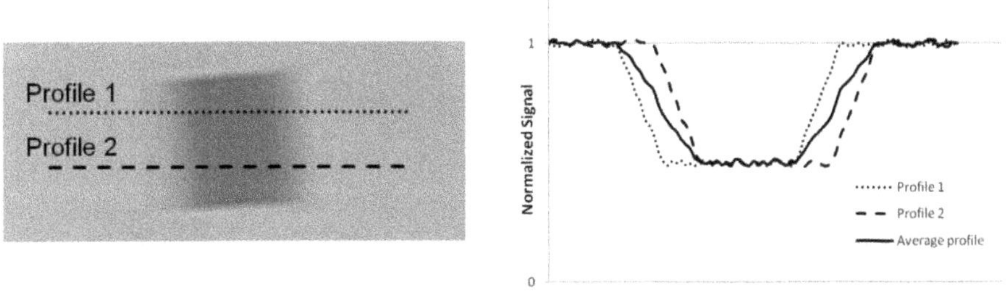

Figure 5.8 Averaging of multiple profiles across a ramp with small in-plane rotation. Noise effects are reduced without compromising the accuracy of FWHM measurement, since: $FWHM_{average} \approx FWHM_1 \approx FWHM_2$

For phantoms with oppositely inclined ramps, two FWHM width measurements, d_1 and d_2, may be made from a single acquired image. These may be combined to give an estimate of slice width, W, which compensates for the acute effects of rotational misalignment about the y-axis (AAPM, 2010):

$$W = \frac{(d_1 + d_2)\cos\phi + \sqrt{(d_1 + d_2)^2 \cos^2\phi + 4 d_1 d_2 \sin^2\phi}}{2\sin\phi} \tag{5.7}$$

where ϕ is the angle between the ramps' inclined surfaces (i.e. $\phi = 180\text{–}2\theta$). Assuming careful alignment of the phantom, a good approximation of Equation 5.7 may be obtained by separately inserting d_1 and d_2 into Equation 5.6 and taking the geometric mean of the resultant values of W.

Accurate measurement of slice width may be compromised if there is substantial noise in the image. Noise in the signal profile may be reduced by averaging multiple ramp profiles from a single image. Unlike wedge ERFs, measurement of an accurate FWHM from averaged profiles is not compromised by small in-plane rotational misalignments of the phantom, therefore co-registration of profiles before averaging is not necessary (Figure 5.8).

Accurate measurement of slice width may also be compromised if there is substantial non-uniformity in the image due to, for example, increased signal in close proximity to coil elements in unfiltered images. Figure 5.9 illustrates how such non-uniformity may affect the signal profile. Attempts to measure the FWHM either manually or using curve-fitting may benefit from the application of a uniformity correction filter. It may also be beneficial to avoid orienting the ramps parallel to the z-axis in sagittal and coronal acquisitions, as signal non-uniformity is commonly most conspicuous on this axis.

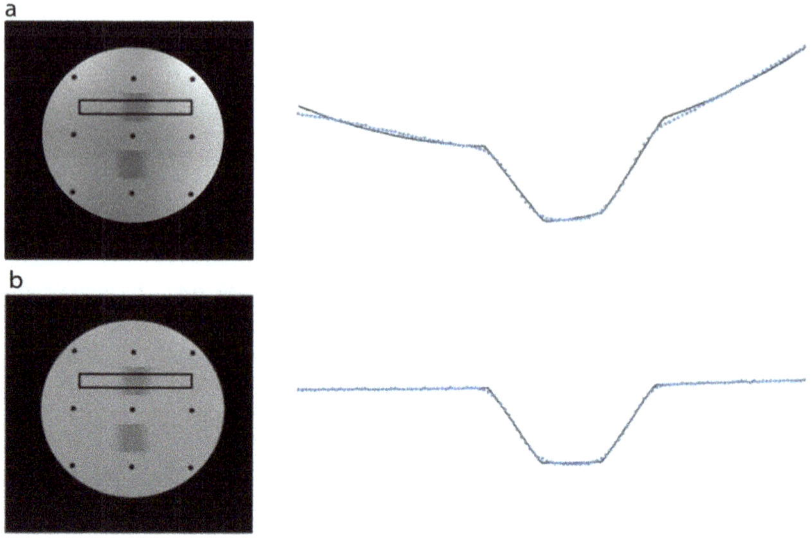

Figure 5.9 Ramp profiles from (a) unfiltered and (b) uniformity corrected sagittal head coil images. In regions of pronounced background non-uniformity, attempts to model the profile with a trapezoidal function require addition of a function to account for non-uniformity. In this instance, the solid lines represent the best-fitting sum of a trapezoidal function and a second-order polynomial

For hot ramps, the profiles will be inverted. It should be noted that in magnitude acquisitions with low SNR, rectified noise in non-signal-producing regions might affect FWHM measurements.

Determining slice width using a ramp imposes a restriction on the measurement of thin slices due to the finite thickness of the ramp itself (Figure 5.10).

It should be noted that when using ramp test objects, the signal profile obtained from the image is **not** the slice profile as defined in Section 5.2.2. Rather it is the slice profile convolved with the two-dimensional cross-section of the ramp's intersection with the acquired slice. Precise determination of the slice profile requires deconvolution of the signal profile, which is not detailed here. Alternatively, if the ramp thickness is significantly less than the slice thickness (NEMA (2010a) suggests that it be less than one-fifth of the selected slice thickness times the inclination angle cosine), the signal profile from the image gives a close approximation of the slice profile.

5.4.2.2 Measurement of slice position

Slice position accuracy may be assessed using a similar approach to that using wedges described in Section 5.4.1. In this case, *s* is the relative shift between ramp profiles in the two acquisitions.

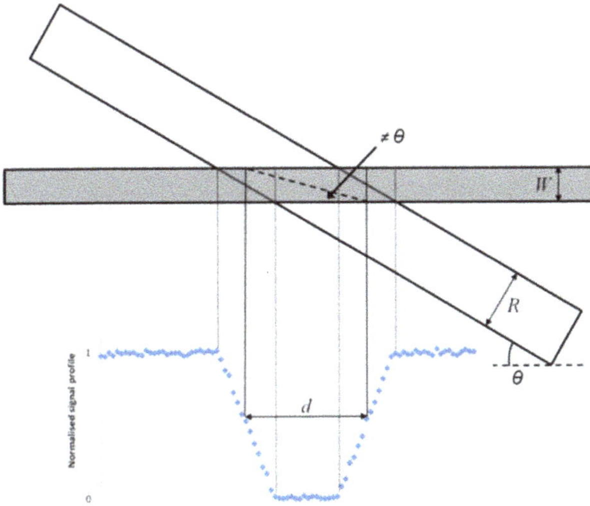

Figure 5.10 In this acquisition, the centre of the profile falls to zero where the cold ramp occupies the full thickness of the acquired slice. The relationship between the FWHM of the signal profile, d, and the slice width, W, no longer holds. In fact, the FWHM is now dependent on R, the physical thickness of the ramp itself: $d=R/\sin\theta$. The slice width W can only be measured using a ramp if: $W \geqslant R/\cos\theta$

5.4.3 Rod methods

5.4.3.1 Measurement of slice position

Figure 5.11 illustrates acquisition of an axial slice through cold crossed rods (with a relative inclination of θ) immersed in signal-producing fluid. The slice is displaced by a distance P from the reference plane, at which the rods are in closest proximity.

The measured slice position, P, is given by:

$$P = \frac{s}{2\tan\left(\frac{\theta}{2}\right)} \tag{5.8}$$

For phantoms in which the rods cross at right angles, this is simplified to:

$$P = \frac{s}{2} \tag{5.9}$$

It is recommended that slice position be measured at multiple locations by performing a multi-slice acquisition or a 3D volumetric scan. This allows assessment of performance over the full field of view facilitated by the phantom (Figure 5.12), and enables the precise location of the reference plane to be determined (using interpolation if necessary), rather

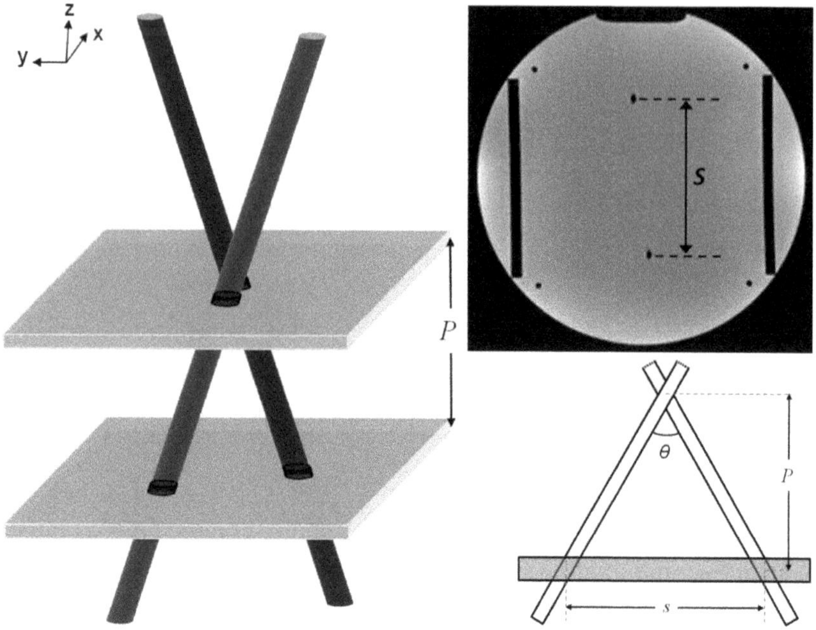

Figure 5.11 Axial acquisition through cold crossed rods at a displacement P from the reference plane. The rod separation, s, is the ***orthogonal*** *displacement between the centres of the imaged rods*

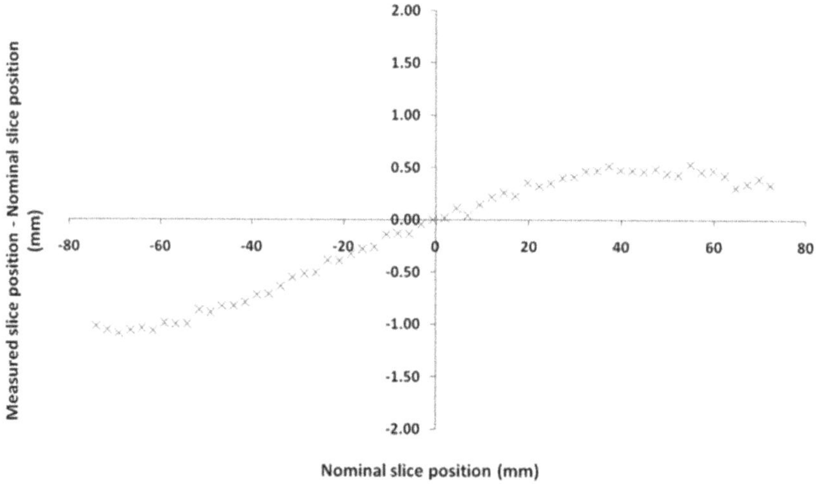

Figure 5.12 Typical errors between measured and nominal slice positions from a 60-slice axial 2D acquisition

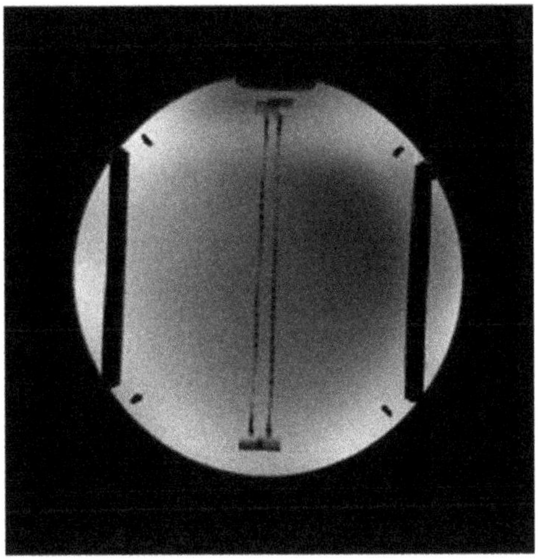

Figure 5.13 A through-plane minimum intensity projection of a 60-slice 2D acquisition into which a slight rotational misalignment has been introduced. The central parallel 'tracks' depict the superposition of the rods' locations in each slice. The effects of in-plane rotation may be compensated by dividing measurements of s by the cosine of the misalignment angle

than relying on the laser positioning system. Furthermore, it enables measurements of s to be corrected for any in-plane rotational misalignment (Figure 5.13). For 3D acquisitions, a 3D distortion correction filter should be applied where available.

Some commercial phantoms contain multiple crossed rods (e.g. Figure 5.3d). This allows characterisation of slice warping – the extent to which slice position varies across the imaging plane (Lerski, 2003).

5.5 Measurement of slice parameters

5.5.1 Purpose and frequency of measurement

The purpose of this measurement is to quantify the error between measured and prescribed slice parameters. These tests should form part of acceptance testing and pulse sequence evaluation and should be performed as part of annual QC procedure or in the event of changes to the hardware or a major system upgrade. For sites using MRI in applications where geometric accuracy is critical, such as radiotherapy treatment planning, stereotactic surgery or clinical trials involving quantitative measures, testing should be performed at frequent intervals on the specific pulse sequences used clinically.

Table 5.1 Relative merits of phantom types

Phantom type	Advantages	Disadvantages
Wedge	Suitable for measurement of slice width, profile and position No lower limit on measurable slice width	Accurate slice width measurement may require a differentiation step Averaging multiple profiles from a single image may require compensation for in-plane rotational misalignment of phantom
Ramp	Most straightforward method of measuring slice width Multiple profiles from a single image may be averaged directly	Useful slice profile characterisation only possible with a very thin ramp (for which the SNR will be low)
Rod	Phantoms commercially available for slice position measurement over a large volume	Slice width and profile measurement not straightforward due to the circular cross-section of the test object

5.5.2 Phantom selection

The relative merits of the phantom types discussed are summarised in Table 5.1. It is recommended that phantoms containing two crossed test objects be used to minimise gross errors due to positioning inaccuracies. Phantoms with broader wedge or ramp test objects are more amenable to improving SNR by taking multiple profiles from a single image. Multiple test objects within a single phantom may permit characterisation of slice warping. The overall size of the phantom may restrict its use with certain coils or in certain orientations.

5.5.3 Procedure

For routine quality assurance purposes, it is recommended that a number of the manufacturer's fundamental sequences be tested. This approach allows a range of RF transmission pulse schemes employed by the scanner to be investigated. These should include 2D spin echo and 2D gradient echo, with consideration given to incorporating 2D fast spin echo and 3D spoiled gradient echo. For 3D acquisitions, the effective slice profile is dependent on partition encoding rather than the slice selection pulse. Selection of a "low SAR" RF pulse type will fundamentally affect the slice profile; such pulses should be tested separately if it is considered appropriate.

When performing the acquisitions, additional imaging and reconstruction options such as phase oversampling, flow compensation, zero interpolation filling (zero padding) and parallel imaging should be turned off. Distortion and uniformity correction filters should remain enabled to aid measurement accuracy, and to differentiate slice selection inaccuracies from other measurable effects (see Section 5.4.2). The receiver bandwidth should be fixed to ensure replication of testing conditions between surveys.

Typically, head coils are used as they permit high-SNR images with the facility for good uniformity correction. Since slice selection is dependent on well-calibrated and linear gradients, testing should be performed in each of the principal planes where practicable. The choice of specific acquisition parameters should be informed by both the geometry of the test object and the relaxation properties of the signal-producing medium.

- Clinically relevant slice widths should be used (e.g. 5 mm and 3 mm for 2D sequences). If ramp test objects are used, there is a lower bound on the measurable slice thickness (see Section 5.4.2) dependent on the physical dimensions of the ramp itself.
- For 2D scans to determine slice width, a single slice may be acquired. Considerations for slice position measurements are detailed in Sections 5.4.1 to 5.4.3.
- TR should be sufficiently long to allow full relaxation (at least three times the T_1 of the signal-producing medium).
- Echo time (TE) should be short enough to minimise signal loss due to T_2 relaxation, but not so short as to restrict the RF pulse duration such that slice profile is compromised. Typically, a TE of 30 ms is used.
- In-plane spatial resolution of approximately 1 mm is sufficient in most cases (e.g. a 256 $\times$ 256 matrix covering a 250 mm $\times$ 250 mm field of view). For thinner slices or steeply angled test objects, higher in-plane resolution may be necessary to obtain useful metrics.
- If slice width measurements are compromised by low SNR, it may be necessary to increase the number of signal averages (NSA). Alternatively, multiple profiles from a single image may be averaged. This approach should be employed with caution if wedge phantoms are used (see Section 5.4.1).

Phantoms should be placed as close as possible to the coil centre and magnet isocentre. Precise orthogonal alignment is critical for single-wedge or single-ramp phantoms, and less so for those with two oppositely inclined edges. Localiser/scout scans should be used to ensure appropriate positioning accuracy of the phantom.

5.5.4 Analysis

The analysis of slice width and position is carried out as outlined in Section 5.3 for the chosen method. For slice width profiles, it is best to average several lines to improve the profile SNR. When calculating the FWHM, it may be necessary to take account of non-linearity of the MR images and apply some detrending to the profile to accurately assess the baseline. Some pixel averaging may also be necessary to reproducibly ascertain the true maximum height.

5.5.5 Pitfalls

MR image non-uniformity can result in errors in the baseline estimation. It may be prudent to use a more uniform coil, e.g. a birdcage coil rather than an array coil. Some scanners may have a T/R head coil, but if not, then the inbuilt body coil may produce a more uniform response. Some additional signal averaging may be required to make up the loss in SNR.

Very thin slices cannot be investigated using ramp methods due to the under-sampling introduced by the image matrix and the ramp thickness. The resultant profile will be convolved with both. As a rule of thumb, the nominal slice should not be less than three times the ramp. For very thin slices (or 3D LSF), the wedge method is more accurate. It does however require a high image SNR.

Finally, it is important to be aware of the effect of the phantom filling material's relaxation times, particularly T_1. The value of TR chosen should not be less than $3 \times T_1$. Gradient-echo sequences are particularly likely to produce a distorted profile (McRobbie *et al.*, 1986).

5.6 Interpretation

5.6.1 Slice profile and slice width

If the slice profile is poor, it affects the detectability of small objects such that lesions coincident with the edge of the slice will have poorer contrast than similar lesions coincident with the centre of the slice. It may also result in significant cross-talk in multi-slice acquisitions (with a consequent drop in SNR and detection sensitivity), although this effect may be mitigated by increasing the slice gap or using interleaved slice acquisition schemes. If the measured slice width differs significantly from the nominal value, the desired detection sensitivity of the sequence is fundamentally altered, again with consequences for SNR.

Each slice width measurement should be expected to lie within 10% of the nominal slice width for QC acquisitions (a 14% tolerance is stipulated as part of the ACR accreditation programme (ACR, 2005)). Slice width measurement errors may arise from poor phantom positioning or inappropriate sequence design (see Section 5.5.3); inappropriate averaging of profiles; interpolation errors; or due to uncertainty introduced by significant noise or signal non-uniformity. Sources of disagreement between nominal and measured slice width include RF pulse distortion (particularly as a result of RF amplifier non-linearity) or instability (AAPM, 2010); B_1 field inhomogeneity; cross-talk (where a multi-slice acquisition is performed); ineffective spoiling of stimulated echoes; significant miscalibration or non-linearity of the slice selection gradient.

5.6.2 Slice position

If the measured slice position differs significantly from the nominal value, the accuracy of distances measured along the axis under test is compromised. This is of particular concern for clinical applications where geometric accuracy is integral, such as radiotherapy treatment planning. Measured slice position should be expected to lie within 1 mm of the nominal slice position. Slice position measurement errors may arise from failure to account for errors in phantom positioning (see Section 5.4.3); inappropriate sequence design (see Section 5.5.3); or unaccounted-for automated table movement between sequences (ACR, 2005). Sources of disagreement between nominal and measured slice position include miscalibration or non-linearity of the slice selection gradient; significant B_0 field inhomogeneity (and an ineffective distortion correction filter if applied); or a faulty table positioning mechanism. Depending on system vendor, multi-slice acquisitions utilise either table positioning or the slice-select frequency offset to define slice position; should an error be observed, it is important to know which system is in effect.

If significant disagreement is found in measurements of either slice width or slice position, corrective action by the service engineer is merited.

6 Relaxation times and contrast: T_1, T_2, and T_2*

John Thornton

6.1 Introduction

In MRI, the macroscopic nuclear magnetisation returns towards equilibrium following excitation or inversion by the exponential processes of spin–lattice and spin–spin relaxation, with time constants T_1 and T_2, respectively. When spin–spin relaxation is combined with the unfocused effects of magnetic field inhomogeneity, transverse relaxation is governed by the time constant T_2*. Since these processes are sensitive to water mobility and concentration, as well as the availability of relaxation centres such as macromolecules or cellular debris, and tissue magnetic susceptibility, all of which may change in soft-tissue pathology, T_1, T_2 and T_2* have provided the basis of contrast in medical MRI since its inception.

Images with contrast weighting in terms of T_1, T_2 or T_2* are acquired in virtually all clinical MRI examinations: the accuracy of the relaxation-time contrast weighting in routine diagnostic MRI, and the stability of this contrast are clearly important. The signal-parameter related QC procedures recommended in Chapter 3 provide measures to detect instrumental defects impacting upon relaxation-time dependent image contrast, and an additional procedure to explicitly monitor relaxation-dependent image contrast is suggested at the end of this chapter. In many non-specialist clinical MR units where MR relaxation time measurements are not performed, these relaxation-time related QC procedures may be considered sufficient.

The majority of this chapter concerns assessing the relaxation time **measurement**, or 'relaxometry', performance of the MRI instrument. Image-based T_1, T_2 or T_2* quantification is an endeavour distinct to the ubiquitous generation of relaxation-time weighted diagnostic images, and requires more sophisticated acquisition, data analysis and quality control methods. Following a brief outline of the clinical and research importance of MR relaxometry, the specification and design of appropriate test objects, and methods for their calibration are described. The use of these test objects in baseline and ongoing routine relaxometry QC are then discussed before the chapter ends with a description of relaxation-time weighted contrast QC.

6.2 MR relaxometry

Clinically, MRI relaxometry to inform patient management, or to provide constituent data for more complex functional analyses, is becoming increasingly widespread. While space does not permit a full review, clinical applications include T_2/T_2* assessment of iron overload (Wood, 2007) in the liver (St. Pierre *et al.*, 2005; Echeverría *et al.*, 2012) or heart (Kondur *et al.*, 2009; Pennell, 2005); T_2 mapping to characterise tumours in the brain (Hattingen *et al.*, 2013; Wagnerova *et al.*, 2012) or prostate (Roebuck *et al.*, 2009), T_1 as an index of liver cirrhosis (Heye *et al.*, 2012) or T_2 assessment of neonatal encephalopathy (Shanmugalingam *et al.*, 2006). Relaxometry may also represent a necessary step in more complex MRI-based physiological measurements, e.g. T_1 maps required as a component of myocardial extracellular volume measurement (Kellman *et al.*, 2012), arterial spin labelling cerebral perfusion measurement (Deibler *et al.*, 2008) or dynamic contrast enhanced membrane permeability measurement (Tofts and Parker, 2013). It is likely that new clinical relaxometry applications will emerge during the lifetime of this report.

Examples of human *in vivo* MRI relaxometry use in research are numerous; for example, brain relaxometry (Deoni, 2010) has found application from early maturation (Sled and Nossin-Manor, 2013) to aging (Callaghan *et al.*, 2014), offering potential markers of disease progression and response to therapy (Du *et al.*, 2011; Hasan *et al.*, 2012; Walsh *et al.*, 2014).

6.3 In vitro versus in vivo QC

This chapter is focused upon test-object-based QC to detect instrumental or environmental fault conditions. It is recognised however that, under appropriate circumstances, longitudinal *in vivo* measurements may offer complementary measures of measurement stability and precision (Tofts and Collins, 2011); in healthy human subjects, the requirements discussed in Section 6.4.1 for test-object relaxation properties are clearly automatically satisfied, and thermoregulation in healthy individuals avoids the challenges of thermal stability discussed in Section 6.4.3. *In vivo* QC measurements may also detect variations caused by errors in subject positioning or scan geometric prescription. However, test-object QC is likely to offer more sensitive, earlier detection of instrumental faults than *in vivo* QC, which, outside of structured clinical research studies, presents potential ethical and logistical difficulties.

6.4 Test object design

QC requires test objects with appropriate, known, stable MR relaxation times, and physical dimensions appropriate to the MRI equipment employed *in vivo*. Suitable items may be available for purchase commercially, or constructed in-house, according to the principles outlined in Chapter 2.

6.4.1 T_1 and T_2 values

In vivo, T_1 and T_2 vary with tissue type, pathology and field strength as exemplified in Table 2.1 in Chapter 2. A properly designed relaxometry pulse sequence uses timing parameters: echo time (TE), repetition time (TR) and inversion times (TIs), optimally matched to the expected *in vivo* relaxation time range (e.g. Shrager *et al.*,

1998). Failure to achieve this compromises accuracy and precision. Since the QC procedure should ideally employ identical or similar sequence parameters to those employed *in vivo*, a corollary of the above is that the test object should exhibit relaxation properties appropriate to these timings. For this reason, the most readily available test substance, distilled water, is likely unsuitable due to its long relaxation times (estimates vary and depend upon paramagnetic solutes, but typical values at 1.5 T are 3388 ms for T_1 at 23°C (Tofts *et al.*, 2008) and 1370 ms for T_2 at 22°C (Tofts *et al.*, 2000)) compared with most biological tissues. An ideal test object might comprise a number of compartments with a range of T_1 and T_2 values encompassing those expected *in vivo*. Clearly the range chosen will depend upon the field strength and anatomical target, and the practicalities of manufacture.

Chapter 2 provides guidance on test object media and control of T_1 and T_2; since matching these precisely and simultaneously for specific tissues is challenging, a degree of compromise will be necessary. Nevertheless, even with simple paramagnetic salt solutions, values in the range previously recommended by the AAPM (Price *et al.*, 1990), 100 ms $< T_1 <$ 1200 ms and 50 ms $< T_2 <$ 400 ms, should be attainable.

6.4.2 Simulating tissue T_2*

Specialist phantoms are required for T_2* relaxometry QC. In the simplest case, aqueous solutions or gels containing paramagnetic salt may suffice, providing rather long T_2* values close to T_2 in a well-shimmed magnet. A number of reports include calibration data relating T_2* to $MnCl_2$ concentration at various field strengths (Meng and Lei, 2008; Preibisch *et al.*, 2008; Yang *et al.*, 2010). A more sophisticated approach is to introduce microscopic magnetic susceptibility variations by suspending in a gel medium super-paramagnetic iron oxide particles (SPIOs), which influence T_2* according to field strength and particle size (Sedlacik *et al.*, 2014). Phantom designs with corresponding T_2* values at 1.5 T (Ernst *et al.*, 2013; Hines *et al.*, 2009) and 7 T (Hezel *et al.*, 2012) have been described. Potential sources of SPIOs for phantom fabrication are commercially available MR contrast agents approved for imaging of the liver (Wang *et al.*, 2011).

6.4.3 Thermal stability

Both T_1 and T_2 in general are temperature-dependent. While judicious choice of phantom materials may minimise this dependence, temperature stability cannot be assumed, with typical test-object media T_1 and T_2 increasing approximately linearly with temperature with coefficients between 1 and 3% °C^{-1} (Nelson and Tung, 1987). It has also been shown that phantom T_2* may be temperature-dependent, with a reported 1.5% °C^{-1} temperature coefficient in one set-up (He *et al.*, 2009). Sophisticated designs for temperature-controlled MRI test objects have been proposed using heat-bath water-controlled temperature (Xin *et al.*, 2012; Reischauer *et al.*, 2009) or temperature control based on set-point (freezing-temperature) references

(Chenevert *et al.*, 2011; Vescovo *et al.*, 2013). An alternative strategy is to control test-object temperature passively by surrounding a conventional phantom with efficient thermal insulation (Tofts and Collins, 2011), maintaining a temperature previously obtained in thermal equilibrium with the ambient room temperature or that of a water bath.

A more common scenario is to use a non-thermally insulated test object which should reside within the scanner room for a sufficient time before measurement to attain thermal equilibrium. Without precautions to control the phantom temperature, relaxation times may vary by up to 10% over a plausible range of test-object/scanner bore temperature variation (Orr *et al.*, 2007). Determining the test-object temperature is therefore highly important for proper interpretation of QC data. The phantom temperature should be recorded for phantom calibration, and a log of phantom temperature should be kept along with relaxation time data in routine QC. MRI scanner bore temperature, while ideally constant, may in fact vary significantly due to environmental control system issues or the heat generated during scanning. To account for potential thermal drift, temperatures should be recorded at both the beginning and end of the scanning session.

Phantom temperatures must be recorded with an MR-conditionally safe thermometer; for phantom calibration purposes, fibre-optic temperature sensors, accurate to ± 0.2°C, are ideal and may be introduced directly into the phantom solution/gel with minimal disturbance of the scanner B_0 and B_1 fields (Buchenberg *et al.*, 2015). A thermocouple probe next to a well-insulated phantom can be effective (Tofts and Collins, 2011), provided care is taken to avoid regions of probe-related image artefact in the data analysis. For non-insulated phantoms, surface temperature measurements may suffice, but these will only represent internal temperatures if sufficient time has been allowed to attain thermal equilibrium with the surrounding air. For routine QC measurements in situations where measurement time and complexity are constrained by service practicalities, a readily commercially available self-adhesive strip thermometer attached to the phantom wall may provide sufficient accuracy, but the temperature calibration should be confirmed against a suitable reference.

6.4.4 Chemical stability

In gel phantoms, as *in vivo*, MR relaxation times are strongly dependent upon water concentration; the gel concentration largely determines T_2 (Mitchell *et al.*, 1986), which may therefore drift if the water content decreases by evaporation over time. For this reason, relaxometry phantom compartments must be well sealed. It is also common to incorporate an antibacterial or fungicidal agent, such as sodium azide, to prevent gel degradation due to biological activity. If test objects are to be used over long time scales (years), it is advisable to perform periodic calibration checks (Heng *et al.*, 2014).

6.4.5 Size and physical stability

The test object should fit comfortably within, and ideally fill, the RF receive coil used *in vivo*: this may be particularly challenging for contemporary multi-channel head coils due to their typical tight-fitting, curved designs. It is important that the phantom is sufficiently heavy, or securely fixed, to avoid mechanical vibration while scanning. The low viscosity of water or other liquid-based phantoms means that internal motion may degrade images unless considerable settling time is allowed between the final phantom position adjustment and acquisition; gel-based test objects are particularly advantageous in this respect since the test-object material is relatively immobile.

Phantom size may also influence B_1 homogeneity due to wavelength-dependent dielectric resonances (see Chapter 2, Section 2.3); test objects should therefore consist of compartments small compared with the RF wavelength, or of materials with low dielectric constants. This is important since relaxometry is generally sensitive to B_1-dependent flip-angle variation.

6.5 Test object calibration

6.5.1 Sources of calibration data

For many applications, including determining the accuracy of a newly implemented relaxometry method, or assessing inter-centre reproducibility for a multicentre trial, it is vital that ground-truth test-object relaxation times are known. The most straightforward solution may be to purchase a commercially available test object with a certificate of calibration. However, external calibration may be in error, and relaxation times may in any case drift over time due to gel degradation, necessitating in-house calibration and procedures to monitor long-term stability. If a reliable third-party calibration is not available, then in-house calibration with purpose-optimised pulse sequences and analysis methods should be performed. Results may then be usefully compared with literature data for the specific phantom medium 'recipe', or reference values obtained for the same test object on an independent MR system.

6.5.2 Practical phantom calibration – general caveats

In principle, calibration involves the acquisition of a series of images, systematically varying one key timing parameter (TE, TR, TI), while holding all other experimental conditions constant between each acquisition. Appropriate signal behaviour curves are fitted to numerical values from each pixel or user-defined region of interest (ROI), yielding estimates of the relevant relaxation time. While it is entirely possible to achieve satisfactory calibration using a conventional clinical MRI system with standard, scanner-manufacturer provided pulse sequences, in practice, it is crucial that acquisition and analysis schemes are chosen to minimise systematic measurement errors and maximise precision.

The experimental requirements for phantom calibration are distinct from those of *in vivo* relaxometry: the priority here is accuracy and precision, if necessary, at the expense of measurement time. While preliminary low-resolution pilot measurements may be useful to determine the optimal range of TEs, TRs or TIs to properly capture the exponential relaxation curves, the final calibration measurements may be time consuming since a spatial resolution sufficient to avoid partial volume errors or under-sampling artefacts is required. To avoid parameter-fitting errors, for calibration, all images should yield signal substantially above the noise floor, potentially requiring multiple image averages to boost signal-to-noise ratio (SNR).

It is important that the only condition varying between successive images in a calibration experiment is the intended sequence timing parameter. Automated pre-scan calibration procedures between each image should be avoided to prevent inter-image RF calibration, receiver gain or shim setting variations. Additionally at the analysis stage, any image rescaling applied automatically by the scanner during reconstruction, or by any software used for image file-format conversion, must be properly taken into account.

In conventional 2D MRI sequences, minor out-of-slice RF excitation is common, potentially perturbing the magnetisation within adjacent slices – so-called inter-slice cross-talk. Calibration is thus best performed using single slice acquisition, unless test measurements have explicitly shown that inter-slice cross-talk effects are inconsequential for the sequence concerned.

6.5.3 T_1 calibration recommendations and caveats

There exist many imaging-based methods for measuring T_1 (see, for example, Kingsley, 1999a; Deoni, 2010; Kaldoudi and Williams, 1993), but for all, a potential source of systematic error is spatial variation in the B_1 transmit field. For instance, the well-known simple equation describing signal behaviour in the inversion recovery sequence (S_{IR})

$$S_{IR} = A\left[1 - 2\exp\left(\frac{-TI}{T_1}\right) + \exp\left(\frac{-TR}{T_1}\right)\right] \tag{6.1}$$

with two unknowns A and T_1 (Kaldoudi and Williams, 1993), is derived from the Bloch equations assuming exact 90° and 180° flip angles, respectively, for the excitation, and refocusing and inversion RF pulses. This assumption may often fail in practice: the MRI system pre-scan RF calibration may be inaccurate, B_1 may vary across the object due to transmit coil RF non-uniformity or RF standing wave effects, and the RF slice section process inevitably yields a through-slice distribution of flip angles. In these situations, which are hard to avoid entirely, sequence signal behaviours become more complex and fitting the data to Equation 6.1 will yield inaccurate T_1 values. Accurate values may be obtained by fitting to a more complex model including explicit terms for B_1 at each spin location, sacrificing some fit precision if B_1 is included as a freely floating model parameter, or requiring a separate B_1 mapping acquisition if B_1 is specified as a fixed parameter. If there is a distribution of flip angles across each measurement volume due to, e.g. slice profile variation, then, in general, the solution also involves integration across the slice profile.

These complications may be largely avoided by using a pulse sequence for which the signal behaviour may be fitted to a model implicitly accounting for B_1 variation. For instance, it is straightforward to show (Kingsley, 1999b) that when the excitation, refocusing and inversion flip angles are not accurate, Equation 6.1 may be replaced, respectively, with Equation 6.2, now with three unknown values, where the parameters A and B absorb flip-angle variation effects:

$$S_{IR} = \left[A - B\exp\left(\frac{-TI}{T_1}\right) \right] \tag{6.2}$$

Equation 6.2 holds for inversion recovery data even though T_R is insufficient to allow full relaxation, providing T_R is held constant as TI increases (Gupta *et al.*, 1980), and even when the flip angles vary across the signal generating volume (Evelhoch and Ackerman, 1983), such that the effects of flip angle variation across the slice profile are inherently incorporated. It is therefore recommended that, for T_1 phantom calibration, succesive inversion recovery images are acquired, ideally with $\text{TR} > 5T_1$ to maximise the available SNR, but otherwise holding TR constant as TI increases. It has been shown that fitting inversion recovery imaging data to Equation 6.2 rather than Equation 6.1 can avoid significant underestimation of test-object T_1s (Kingsley *et al.*, 1998).

6.5.4 T_2 calibration recommendations and caveats

T_2 phantom calibration with multiple-echo spin-echo imaging sequences commonly available on clinical MRI systems is also susceptible to in-plane and across-slice B_1-variation-related errors. This causes contamination of the echo-train with unwanted coherences, causing, for example, the first (uncontaminated) echo relative-amplitude to appear lower than the subsequent echoes, and decreasing the rate of signal decay in the remaining echoes (Crawley and Henkelman, 1987; Majumdar *et al.*, 1986; Hennig, 1988). While sophisticated correction schemes have been suggested (Lebel and Wilman, 2010), and for *in vivo* purposes these errors may be manageable (Pell *et al.*, 2006), for phantom calibration this situation is best avoided.

It is recommended therefore that on a clinical scanner, test-object T_2 calibration is confirmed using a simple single-echo spin-echo imaging sequence, acquiring individual images with successively increasing echo times, and TR sufficiently long ($\text{TR} > 5 \times T_1$) to provide essentially complete relaxation between repetitions. In this case, T_2 may be recovered accurately by fitting the image pixel intensities, S_{SE}, according to

$$S_{SE} = A \exp\left(\frac{-TE}{T_2}\right) \tag{6.3}$$

Provided all conditions apart from TE remain constant between the individual echo acquisitions, fitting Equation 6.3 to the signal should return accurate T_2 values despite B_1 variations, the effects of these being absorbed into the amplitude parameter A.

6.5.5 T_2* calibration recommendations and caveats

T_2* calibration measurements can be conveniently performed using a gradient-echo imaging sequence with multiple TEs. Provided again that TR is sufficiently long to allow complete T_1 recovery between excitations, and care is taken to minimise main field B_0 inhomogeneity, T_2* may be estimated by fitting to the image data the theoretical signal intensity, S_{GE}, equation

$$S_{GE} = A \exp\left(\frac{-TE}{T_2^*}\right) \tag{6.4}$$

The most important potential confound is the effect of macroscopic magnetic field variation, due to inadequate B_0 shimming or magnetic susceptibility discontinuities at phantom edges or between phantom compartments. This field variation adds to the intended microscopic susceptibility variation within the phantom material, causing exaggerated signal decay and underestimation of T_2*. While for *in vivo* T_2* relaxometry sophisticated acquisition or post-processing solutions to this problem may be required (Yang *et al.*, 2010), for test-object-based calibration and QC, reliable values may be obtained with simple precautions: first, attention should be given to the scanner pre-scan field homogeneity adjustment, and if possible, the global shim linewidth recorded for reference. Second, a high spatial-resolution 3D acquisition, or narrow slice width for 2D should be used to minimise across-voxel field variation. Third, images should be visually inspected and calibration data extracted for analysis from regions away from phantom internal and external walls or air pockets.

6.5.6 Relaxation time estimation

Once calibration images have been acquired, phantom relaxation times are recovered by fitting appropriate signal models to the data, achieved either on a ROI basis, or pixel-by-pixel to produce relaxation time maps. In the first scenario, ROIs are defined on the first image and propagated identically through the image data set, and the mean signal for each ROI extracted as input to a suitable fitting routine. Care should be taken to place ROIs in uniform image regions free of artefact or partial volume contamination. Non-linear least-squares fitting may be performed using analysis functions within spreadsheet software such as Microsoft

Excel, mathematical programming platforms such as MATLAB (MathWorks, 2015) or Mathematica (Wolfram, 2015), or with specialist scientific data analysis packages such as SigmaPlot (Systat Software Inc., 2015).

The fitting process should generate a graph of the data and fit function for visual assessment of the data and fit quality, a numerical indication of goodness of fit, typically the coefficient of determination (r^2), and, crucially, estimated error-bounds of the fit parameters. The latter give a quantitative measure from the fit of the calibration precision.

Pixel-wise signal fitting to generate relaxation time maps requires specialist image analysis software. This may be provided by the scanner vendor, but greater flexibility is generally obtained with tools written in-house, or within commercial, e.g. Jim (Xinapse, 2016), Analyze (AnalyzeDirect, 2015) or free-to-download packages, e.g. ImageJ (NIH, 2015), MR-map (Messroghli, 2015). It is good practice always to compare results obtained with a particular software with those obtained by re-analysing the same data with a second program, since the possibility of programming errors cannot be completely eliminated, even with commercial software. Once the map is generated, regional mean values may be obtained by defining suitable ROIs, and precision in this case estimated in terms of the ROI standard deviations.

6.6 Relaxometry QC

6.6.1 Purpose and frequency

The purpose of QC is to establish at the outset a method's accuracy and short-term repeatability, and to then monitor variation over time so that instrumental or operator errors may be detected and rectified. Numerous factors may compromise *in vivo* relaxometry, often in ways not immediately apparent to the scanner operator. These include instrumental issues: system timing errors, transmit or receive gain instability, RF coil malfunction, excessive main field inhomogeneity, side effects of software upgrades, and human factors such as incorrect scan prescription during the examination, or inconsistent patient positioning. Detecting and troubleshooting instrumental errors require regularly-acquired, standardised quality assurance data which may be referenced to baseline values.

The frequency at which QC is carried out is dependent upon the clinical (or research) context, e.g. whether the relaxometry is a per-patient study or part of an ongoing longitudinal study, or a multicentre trial.

6.6.2 Procedure

6.6.2.1 Baseline QC

Baseline QC determines reference values against which future routine QC measurements may be compared. If the pulse sequence and parameters used for QC are identical to the *in vivo* protocol, and the phantom calibration relaxation times correspond to the expected *in vivo* range, this amounts to establishing the underlying accuracy of the *in vivo* measurements. Baseline data are obtained by scanning the calibrated phantom with the QC sequence, parameters and RF coil which are to be used for all future QC acquisitions. The baseline measurements should be repeated within the same session, ideally removing and repositioning the phantom between each, a sufficient number of times to allow the mean and standard deviation of the relaxation time for each phantom compartment to be reliably estimated. Comparison of the mean values obtained for each compartment with the phantom calibration reference values allows the method's *accuracy* to be quantified; the respective standard deviations give measures of the short-term *repeatability*, or precision, of the test-object measurements. The latter provides a lower (best-case) bound on the variation that can be expected in long-term routine QC measurements, providing a rationale for setting action criteria.

6.6.2.2 Routine QC

The purpose of routine QC is to monitor the long-term accuracy and reproducibility of the relaxometry method, allowing timely remedial action if instrumental errors are detected. Routine QC should employ the same RF coils, acquisition sequence and parameters, and analysis method as the baseline QC, which will ideally correspond to those used *in vivo*. QC should be performed at regular intervals, with a frequency dependent upon how often *in vivo* measurements are performed, and how severe are the consequences of measurement error. In any case, measurements should be performed sufficiently often for the ordinary variation of QC relaxation time values to be reliably determined, so that deviations outside of these limits can be reliably ascribed to an extraordinary variation requiring remedial intervention; every 2 weeks would seem a reasonable minimum.

6.6.3 Analysis

For each QC measurement, ROIs should be carefully placed upon uniform regions within each phantom compartment, and mean and standard deviation recorded for each relaxation time. If a multiple-compartment test-object is used, for each QC measurement it is useful to plot the estimated relaxation time against the reference values, allowing visual assessment of deviation from the line of unity (Figures 6.1b and 6.2b). After each QC measurement, relaxation time mean and standard deviation values for represented phantom compartments should be added to a continuous plot of regularly recorded values against date, forming a run-chart from which statistical levels of normal variation can be defined according to the principles of statistical process control (SPC) (Wheeler, 1992). *Action criteria* may then be specified defining limits of acceptable measurement variation: values obtained outside of these limits imply remedial action is required.

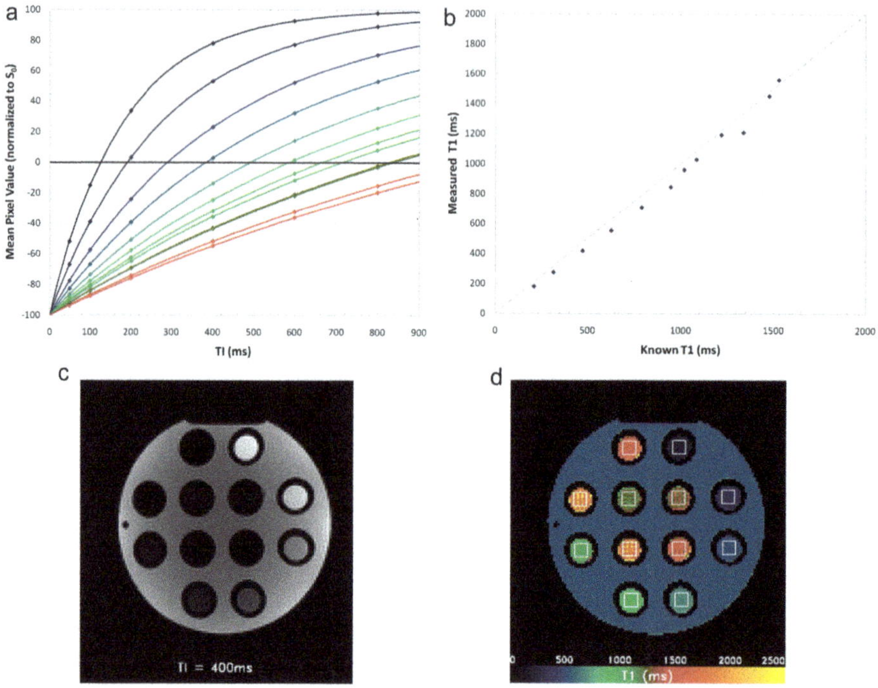

Acquisition parameters			
TR (ms)	1000	Slice width (mm)	5
TE (ms)	15	Receive bandwidth (Hz/px)	130
TI (ms)	50, 100, 200, 400, 600, 800 (separate acquisitions)	NSA	1
FOV (mm)	250 x 250	Single transverse slice	
FOV (px)	128 x 128		

Figure 6.1 Example T_1 routine QC data for the Eurospin relaxometry test object (incorporating 12 gel vials with T_1 values in the range 200–1500 ms at 20 ℃) using multiple spin-echo inversion recovery acquisitions with parameters shown below the figure panels. (a) Inversion recovery curves for each phantom compartment. (b) Comparison of values obtained with reference values. (c) Example raw image from the test dataset. (d) Calculated T_1 map showing analysis regions of interest. Figure kindly provided by Dr Aaron McCann, Radiological Sciences and Imaging, Regional Medical Physics Service, Belfast Health & Social Care Trust

Figure 6.1 shows an example routine QC record for inversion-recovery T_1 relaxometry. For analysis, ROIs were positioned within each phantom compartment so as to avoid partial volume contamination. In this case, to provide a time-efficient QC measurement, the TR was short relative to the maximum T_1 value.

A routine T_2 relaxometry QC measurement is exemplified in Figure 6.2. For the analysis, ROIs are again carefully defined to avoid partial volume contamination. In each phantom compartment, the signal intensity in the image varies with TE to a good approximation according to Equation 6.3, provided, in the case of a multi-echo acquisition such as this, the shortest TE image is excluded from curve fitting. This is necessary because the first TE image signal is a pure spin-echo whereas subsequent TE images contain additional stimulated-echo coherence contributions creating an apparent amplitude inconsistency (Figure 6.2a).

6.6.4 Pitfalls

In all relaxometry QC procedures, certain precautions are necessary to minimise avoidable inter-measurement variation. If QC measurements are performed at the start of the working day, adequate transmitter and gradient system warm-up time from power-up should be allowed to ensure stable measurements. A method for consistent repositioning of the test object between QC sessions should be devised or provided, possibly in the form of a bespoke phantom holder or cradle to minimise variation due to spatial dependencies of the scanner RF transmit and receive intensities, B_0 homogeneity or gradient performance. Ideally, in order to ensure thermal equilibrium with the environment, the test object should be placed in the scanner room for a number of hours before measurement, but in any case, the test object temperature should be recorded at the start and end of the acquisition. A delay sufficient to permit mechanical vibration or liquid motion within the phantom to subside between the final phantom positioning, or final table move, and the QC acquisition must be allowed. To avoid image noise variation, the scanner-room door should be securely closed and all RF coil components securely connected.

Ideally, measurements should use identical pulse sequences, parameters, image geometry and RF coils to the corresponding *in vivo* acquisitions. This is clearly vital for initial validation of measurement accuracy. In practice, for regular QC intended to detect changes in scanner performance, it may be necessary to reduce the scan time by, for example, decreasing TR due to service-limited QC scanner access time. Provided scan parameters are held constant through the baseline and all subsequent routine QC acquisitions, effective QC may still be performed, accepting that, overall, constant systematic differences from ground-truth calibration relaxation time values may be present.

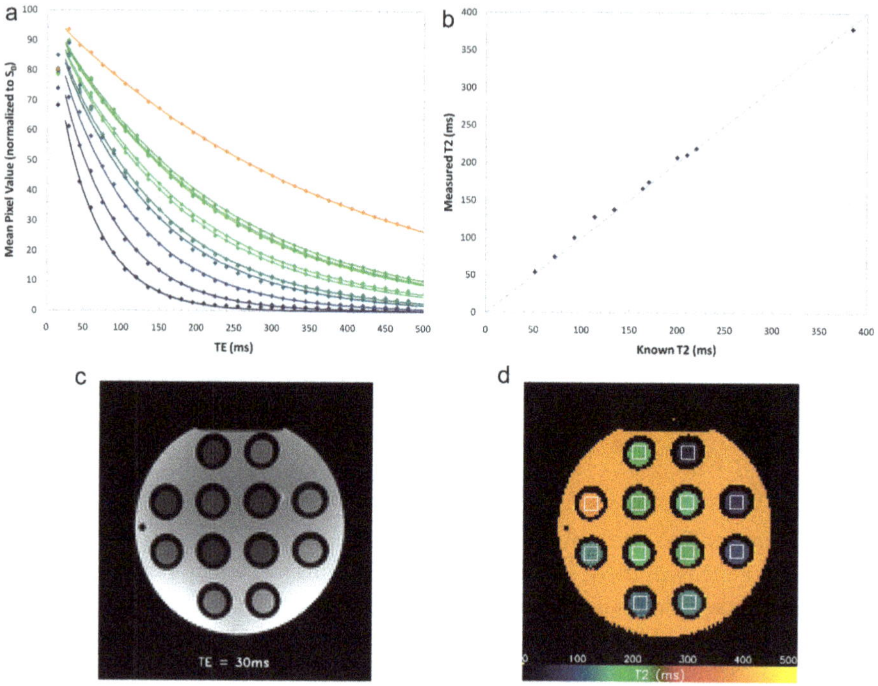

Acquisition parameters:			
TR (ms)	2000	Slice width (mm)	5
TE (ms)	32 echoes; 15ms < TE < 480ms	Receive bandwidth (Hz/px)	130
FOV (mm)	250 x 250	FOV (px)	128 x 128
Single transverse slice		NSA	1

Figure 6.2 Example T_2 routine QC data for the Eurospin relaxometry test object (incorporating 12 gel vials with T_2 values in the range 50–400 ms at 20 °C) using a single multiple spin-echo imaging acquisition with parameters shown below the figure panels. (a) Multi-echo transverse magnetisation decay data for each phantom compartment. (b) Comparison of estimated T_2 values versus phantom-manufacturer provided reference values. (c) Example raw image from the test dataset. (d) Calculated T_2 map showing analysis regions of interest. Figure kindly provided by Dr Aaron McCann, Radiological Sciences and Imaging, Regional Medical Physics Service, Belfast Health & Social Care Trust

6.6.5 Interpretation

Absolute values for action levels depend upon the specific pulse sequence and parameters, phantom and MR system combination, making it impossible to define universal action criteria (AAPM, 2010). Local action criteria are a matter of reasonable judgement: one scheme for individual sample run-charts specifies that further investigation is required for any single QC data point lying more than three standard deviations from the reference value, or whenever two out of three successive QC measurements fall more than two standard deviations from the baseline reference (Oakland, 1996).

If such deviations are detected, then examination of the quality of the theoretical relaxation curve fits to the raw data may provide insight to the source of error. A record of long-term measurement stability will provide evidence of measurement quality where relaxation time measurements are used to support clinical decisions, or investigate long-term longitudinal change in research.

6.7 Relaxation-time-weighted contrast

6.7.1 Quantifying image contrast

In routine radiology, T_1-, T_2- and T_2*-weighted acquisitions are commonly performed for qualitative radiological interpretation rather than measurement. There is nevertheless purpose to quantifying the contrast produced by such sequences, for instance, for imaging protocol optimisation, or for regular QC: fault-related degradation in relaxation parameter-weighted contrast may reduce diagnostic sensitivity.

For the purposes of practical QC, we describe two complementary measures of image contrast. First, the normalised contrast C_{AB} between two uniform regions in an image is defined as

$$C_{AB} = \frac{|S_A - S_B|}{S_A + S_B} \tag{6.5}$$

where S_A and S_B are the mean image intensities across regions A and B, respectively.

A second useful measure is the image contrast-to-noise ratio (CNR_{AB}) defined as

$$CNR_{AB} = \frac{|S_A - S_B|}{\sigma_{AB}} \tag{6.6}$$

where σ_{AB} represents the signal standard deviation, assumed to be common to both regions. Since, in practice, the image signal may in fact vary across nominally uniform phantom regions, σ_{AB} may be most conveniently obtained by measuring the standard deviation $\sigma_{background}$ across a ROI placed in an image background area where no signal is expected. Formally, the background noise in the usual magnitude-reconstructed images follows a Rayleigh rather than Gaussian statistical distribution (Kaufman *et al.*, 1989) and therefore the correction explained in Chapter 3 is required

$$\sigma_{AB} = \frac{\sigma_{background}}{0.665} \tag{6.7}$$

assuming simple magnitude reconstruction with no complications from the use of multi-array receive coils or partially parallel image reconstruction (Dietrich *et al.*, 2007). Care must be taken to avoid placing the background ROI in a region of artefact or where the background noise magnitude may be non-uniform due to filtering during image reconstruction.

6.7.2 Procedure

Image contrast QC requires a test object with two compartments with different known relaxation times, between which the image contrast is measured. Ideally, these should have relaxation times approximating those of tissue types it is desirable to discriminate, e.g. cerebral grey and white matter (see Chapter 2, Table 2.1 for typical values). Contrast QC assessment may be readily achieved within conventional routine QC protocols, since with judicious choice of test object, the necessary data may be obtained from an acquisition used otherwise to assess system SNR (Chapter 3). The test object should be accurately positioned, allowed to mechanically settle before scanning, and its temperature recorded. The imaging sequence to be tested should be run with acquisition parameters identical for each measurement, and identical to those used *in vivo*, with particular attention to the interslice gap.

6.7.3 Analysis

The images should be transferred off-line for analysis and visually inspected for artefacts before continuing. ROIs are placed across uniform signal regions within each of the two test object compartments and the signal mean and standard deviation for each ROI recorded. An image background region outside of the test object and free of artefacts should then be identified, and an ROI placed to determine the standard deviation of the noise signal. Normalised contrast and CNR are then computed according to Equations 6.5, 6.6 and 6.7, and added to the QC record along with the phantom temperature. Before the first routine QC measurement, a baseline assessment of normalised contrast and CNR mean reference values and short-term repeatability should be performed as recommended for relaxometry QC in Section 6.7.1. This will allow action criteria for the contrast QC measures to be defined.

6.7.4 Pitfalls

Pitfalls in relaxation contrast QC are similar to those in relaxometry QC discussed in Section 6.6.4. Spatial variations in receive-coil sensitivity and transmit B_1 must be carefully considered, since these may strongly affect the apparent contrast between ROIs.

6.7.5 Interpretation

Measured ROI values are expected to follow the relaxation behaviour inherent to the pulse sequence used (for example, see Sections 6.5.3, 6.5.4, and 6.5.5). Significant deviations from the expected contrast behaviour may indicate flip angle miscalibration or B_1 non-uniformity. In the latter case, a direct investigation of B_1 may be beneficial (see Section 7.2.2).

6.8 Acknowledgements

The author would like to acknowledge the helpful suggestions of Professor Paul Tofts during the preparation of the final version of this chapter.

7 Specialist QC

Anna Barnes, Glyn Coutts, Peter Gatehouse, Martin Graves, Ian Marshall, Donald McRobbie, Anastasia Papadaki, Scott Semple, and Stephanie Withey

7.1 Introduction

This chapter attempts to address some of the specific QA/QC requirements of some more advanced MR imaging techniques, although this is by no means an exhaustive list, given the breadth and versatility of MRI techniques and applications.

When implementing new MRI applications or techniques, an assessment should be made as to whether these new techniques require any additional quality control. This decision will be influenced by factors such as:

- Will additional hardware be required? This may be third party hardware, such as patient monitoring devices, devices used to interact with the subject (stimulation presentation equipment used in functional brain imaging studies for example) or even acquisitions using advanced MRI systems.
- Will additional levels of quantification be required? If the technique requires specific measurements to be made in order quantify physiological parameters or function, how reproducible are those measurements?
- Will the new technique require specific test objects to be constructed to validate the technique on a regular basis?
- What is the regularity of any additional QC that is required? Is stability of the MRI system a particular requirement? If so, will more regular QC be required for this technique or application?

7.2 B_0 and B_1^+ mapping

7.2.1 B_0 mapping

Knowledge of the static magnetic field (B_0) uniformity is an important part of MRI scanner quality control. The simplest method of spatially mapping the magnetic field is to use a phase difference technique in a large, ideally spherical, non-conducting uniform phantom. This involves acquiring two images with different echo times (TE). Due to the field non-uniformity, there will be a spatial variation in the precessional (Larmor) frequency. The phase accumulated between the two different TEs is proportional to the local frequency. Assume that image $m_1(x,y)$ is acquired at TE_1 and image $m_2(x,y)$ is acquired at TE_2. If both images are complex, then the phase difference can be calculated from,

$$\Delta\phi(x,y) = angle\{m_1(x,y)\cdot m_2^*(x,y)\} \tag{7.1}$$

where the superscript * denotes complex conjugation.

The difference in echo times is given by,

$$\Delta TE = TE_2 - TE_1 \tag{7.2}$$

Therefore, the estimate of the local resonant frequency, i.e. the frequency difference from the nominal Larmor frequency, is given by,

$$\omega(x,y) = \frac{\Delta\phi(x,y)}{\Delta TE} \tag{7.3}$$

where $\omega(x,y)$ is in radians/second.

Of course, the phase difference can exceed $\pm\pi$ in which case it will be necessary to reduce ΔTE or apply some form of phase unwrapping algorithm to the phase difference images.

As a further practical consideration, care should be taken in ensuring that the scanner generates appropriately scaled data to calculate the phase difference images correctly. Whilst most scanners can generate phase images, these are often scaled such that it is not easily possible to calculate phase in the range $\pm\pi$. It is generally easier to calculate the phase from the real (R) and imaginary (I) images, providing they are equally scaled using the relationship,

$$\phi(x,y) = \arctan\frac{I(x,y)}{R(x,y)} \tag{7.4}$$

Care should also be taken when using multi-channel coils as receivers as the complex data may become corrupted during the coil combination. There is an extensive literature on techniques for rapid B_0 mapping in vivo but they all require customised pulse sequences.

The AAPM (2010) has defined tolerance values as <0.5 ppm root-mean-square (RMS) over a 35-cm diameter spherical volume (DSV) for systems used for routine imaging. For systems using echo planar imaging (EPI) sequences and spectroscopy applications, they advocate a tolerance of <0.1 ppm RMS.

7.2.2 B_1^+ mapping

Knowledge of the spatial distribution of the B_1 transmit field, known as B_1^+, is important in many quantitative imaging techniques which rely on accurate knowledge of the RF flip angle. In most cases, B_1^+ mapping is only of interest in vivo where the effect of the body on the flip angle variation is important. However, it may be of interest to investigate the uniformity of the excitation field of particular coil geometry in a uniform phantom. The simplest method of B_1^+ mapping is to use the double angle method (DAM), ideally in a large uniform, non-conducting, phantom. DAM involves the acquisition of two gradient echo images with flip angles, α and 2α. This produces two images with the following signal responses,

$$S_\alpha(x,y) \propto M_0 \sin(\alpha) \tag{7.5}$$

$$S_{2\alpha}(x,y) \propto M_0 \sin(2\alpha) \tag{7.6}$$

Taking the ratio removes the M_0 dependence. Using the relationship that $\sin(2\alpha)=2\cos(\alpha)\sin(\alpha)$

$$\frac{S_{2\alpha}(x,y)}{S_\alpha(x,y)} = \frac{2\cos(\alpha)\sin(\alpha)}{\sin(\alpha)} \tag{7.7}$$

$$\alpha(x,y) = \arccos\left(\frac{S_{2\alpha}(x,y)}{S_\alpha(x,y)}\right) \tag{7.8}$$

This is valid for $30^\circ < \alpha < 60^\circ$.

The limitation of this approach is that the longitudinal magnetisation must recover nearly to equilibrium or there will be different T_1 weightings in the two images. Therefore, this approach only works if the TR is greater than five times the T_1 of the phantom. Other approaches have been developed to overcome this limitation but they require pulse sequence modifications. There is an extensive literature on techniques for rapid B_1^+ mapping in vivo but they all require customised pulse sequences.

7.3 Diffusion imaging

7.3.1 Introduction

Quantitative measures from diffusion imaging such as apparent diffusion coefficient (ADC), fractional anisotropy (FA) and mean diffusivity (MD) may be significantly affected by technical issues which are specific to commercial vendors. MR centres that are involved in diffusion-weighted imaging would therefore be recommended to employ independent regular QC to ensure the robustness of fundamental diffusion-weighted properties, such as ADC/FA/MD measurements.

As with all QC protocols suggested in this report, phantoms should be constructed which approximately mimic the physical properties being measured in the relevant clinical protocol. The protocol should therefore aim to ensure that the system is able to reliably quantify diffusion parameters within ranges routinely obtained in tissues.

Although increasingly being employed in other body areas, the most common application for diffusion imaging is still in brain where ADC ranges between 60 and 105×10^{-5} mm^2 s^{-1} in white matter, between 60 and 83×10^{-5} mm^2 s^{-1} in grey matter, and between 240 and 440×10^{-5} mm^2 s^{-1} in cerebrospinal fluid (Pierpaoli *et al.*, 1996; Sener, 2001). Suitable phantoms should be used that mimic these diffusion properties.

7.3.2 Materials

Test objects used for diffusion-weighted QC should be constructed from suitable isotropic materials which mimic the diffusion and relaxation properties of tissue. Differing approaches towards ADC phantom construction have been successfully used in the past.

7.3.2.1 Alkanes

Tofts *et al.* (2000) suggested the use of three cyclic alkanes as filling materials for ADC phantoms to mimic tissue ADC properties (cyclooctane, cycloheptane and cyclohexane). At room temperature, these alkanes effectively mimic the ADC properties of brain tissue. This approach has been adopted by other groups, and has proven useful for multicentre QC due to the stability and isotropic nature of the alkanes. Zhu *et al.* (2011) demonstrated significant differences in diffusion measures using these alkane phantoms across multiple sites due to differences in hardware/software used in diffusion imaging. Alkanes, however, may be rather expensive and hazardous to handle.

7.3.2.2 Gels

Laubach *et al.* (1998) suggest the use of agar gels and sucrose solution due to their low viscosity giving them particularly stable measures of ADC due to reduction in any bulk motion during handling. Longer term, gel-constructed phantoms may be prone to degradation (oxidation, etc.) unless stored under stringent conditions.

7.3.2.3 Aqueous solutions

Delakis *et al.* (2004) used aqueous filled test objects to perform ADC QC measurements, using distilled water doped with $CuSO_4$ and a second solution of distilled water doped with $CuSO_4$ and sucrose. The $CuSO_4$ and sucrose solution benefits from $CuSO_4$ reducing the solution's T_2 value to clinically relevant levels (see Chapter 2) and sucrose at high levels (400 g L^{-1}) reducing the mobility of the water molecules to mimic tissue diffusion properties.

Wang *et al.* (2011) proposed the use of the widely available American College of Radiology head phantom for QC of diffusion. Although this phantom does not mimic tissue diffusion properties, this approach benefits from the use of a commercially available test object, negating the need to 'build' a QC phantom and also therefore comes with manufacturer guarantee of phantom construction quality. It is possible to make repeated measures of diffusion properties in commercially available phantoms, although these will not directly assess user ability to measure diffusion properties in normal clinical ranges.

Regardless of choice of filling material, routine measures of diffusion parameters should be repeated and logged over time to determine if any systematic changes occur over time.

7.3.3 Imaging protocols

A QC imaging protocol should be used which mimics the clinical protocol being used in the centre. The same imaging coil, sequence type, spatial resolution and slice thickness should be used for QC as in routine clinical diffusion imaging.

If EPI-based diffusion imaging is performed, then some assessment of EPI image QC should also be performed. Comparison of EPI-related distortion to some gold standard sequence (a spin-echo image for example) should be performed to determine if there are any issues relating to EPI non-linear image distortion that may affect any diffusion imaging measures being made from these data. A distortion phantom (see Chapter 4) may be used to assess any distortion across the image using the EPI-based diffusion sequence. Specific image distortion caused by eddy currents induced directly by application of diffusion-encoding gradients may also be assessed in a distortion phantom by comparison of image distortion in $b=0$ to image distortion in high-b value images (Wang *et al.*, 2011). Additionally, the quality of the fat suppression used in EPI may be addressed using a suitable phantom (Winfield *et al.*, 2014).

7.3.4 Phantom handling

If phantoms filled with aqueous solution are being used for diffusion QC, large-scale bulk motion of the solution during phantom handling and positioning can have significant effects on any diffusion measurements made during the QC protocol. Therefore, diffusion phantoms should be allowed to 'settle' for at least 5 min after being positioned in the centre of the bore for imaging. This should allow any turbulent motion of the filling solution induced during positioning to dissipate.

Temperature will also affect diffusion characteristics in test objects so a log of phantom temperature during QC data acquisition should also be made.

7.3.5 Advanced diffusion measurements

More advanced diffusion phantoms have been suggested by some groups to perform more advanced diffusion QC, such as test objects containing structured fibres in order to perform tractography QC. Test objects have been constructed by research groups for this purpose using a range of types of artificial fibres (Pullens *et al.*, 2010). The range of fibre types (hydrophobic, hydrophilic) and choice of binding (shrink-wrapping of fibres, for example) may be manipulated to alter FA measurements or mimic tissue fibre properties. An in-depth discussion of fibre-tracking phantom construction is outwith the scope of this report.

7.4 QC for BOLD fMRI

7.4.1 Introduction

Functional MRI (fMRI) is based on the blood oxygenation level dependent (BOLD) technique, where local changes in cerebral neuronal activity give rise to signal intensity changes in MRI. A series of T_2*-weighted EPI images are collected during repetitive periods of contrasting tasks (e.g. motion and rest or a verbal task), known as the paradigm. A haemodynamic model is applied to the paradigm and the image data are interrogated pixel-by-pixel to identify this pattern of signal change regionally. As these signal changes are only of a few percent for sensorimotor activation (e.g. 2–3% for 1.5 T, 4–7% for 3 T) or less for cognitive tasks, statistical determinants are used to distinguish areas of activation. The resultant activation data are superimposed on anatomical brain images as maps of statistical metrics. A number of statistical approaches are available, including *t*-tests, correlation, Fourier, wavelet and independent components analyses. See Jezzard *et al.* (2001) for a review of methods and applications.

In research, fMRI subject responses are often grouped together for analysis, based upon an anatomical template. The main clinical application of fMRI is for pre-surgical planning: for example, guiding neurosurgeons with regard to the best surgical approach and maximising tumour resection while preserving basic functions, such as motor function following surgical intervention, or for determining language lateralisation before epilepsy surgery. In clinical fMRI, the patient is normally scanned in a single session fMRI which is assumed to provide accurate information on the subject's functional neuroanatomy. The possible sources of variability observed in fMRI data of the same subject are presented in Table 7.1. Note that QC is required for the MR system and for the acquired human dataset.

7.4.2 Review of methods

An fMRI QC protocol should be easy to implement, relatively quick (no more than 5–10 min) and use scanning sequence and parameters based on those used in clinical practice. Because the fMRI datasets are large, analysis of the QC data should ideally be fully automated and it should include measures that detect changes/problems with the MR system and alert the user.

A homogeneous spherical phantom can be used for the fMRI protocol. Gel phantoms have the advantage of not requiring settling time and are less influenced by scanner vibrations. However, consideration should be given to possible dielectric effects, particularly at higher fields (see Sections 2.3 and 8.4.6). A long time series of EPI images should be acquired, in the range of 100–200 volumes and the first 2–5 volumes should be discarded from analysis to allow signal equilibrium to be achieved.

Table 7.1 Factors affecting fMRI variability and quality

Causes of variability	**Solution**
Physiological variation: • Cardiac and respiratory motion • Flow artefacts	Monitored and filtered or modelled
Scanner hardware characteristics: • Shim, noise variation, drift, stability	Assessed by scanner QC
Subject motion: • Patient movement • Involuntary motion • Stimulus-related motion	Image registration software
Subject position	Normalisation software
Non-specific physiological effects • Level of arousal/degree of cooperation	Not controlled or corrected

A fMRI QC protocol should include measurements of scanner stability (short- and long-term), SNR (Section 3.3) and ghosting (Section 3.6). Weekly tests would show long-term stability and identify hardware problems. A simple way of identifying problems is to apply a series of statistical tests to the long-term measurements of each parameter based on Shewhart charting, where statistical tests compare the latest values measured with the mean and standard deviation (SD) of a baseline, typically comprising at least 20 previous stable measurements (Simmons *et al.*, 1999).

Scanner instability has been addressed by Weisskoff (1996) by comparing the signal variance (or standard deviation) across the time series against the individual image signal variance (related to the SNR). The SNR over the time series, SNR_0, is defined as:

$$SNR_0 = \frac{\sum_{i=1}^{N} m_i}{1.53 \sum_{i=1}^{N} \sigma_i} \tag{7.9}$$

where m_i is the mean signal intensity in a large ROI within the phantom for the ith image and σ_i is the background standard deviation.

The time series fluctuations, F_n, are defined for an $n \times n$ ROI as

$$F_n = \frac{\sqrt{\frac{1}{n-1} \sum_i (m_i^n - \bar{m}^n)^2}}{\bar{m}^n} \tag{7.10}$$

where m_i^n is the mean of the $n \times n$ ROI on the ith time series image (from a total of N images) and the mean signal intensity over all of the series is

$$\bar{m}^n = \frac{1}{N}\sum_{i=1}^{N} m_i^n \tag{7.11}$$

If the noise throughout the times series is purely random, then F_n will be inversely proportional to n, thus the ideal (or theoretical) value $F_{n,t}$ is

$$F_{n,t} = \frac{1}{n * SNR_0} \tag{7.12}$$

F_n is computed for different sized ROIs from n=1 to 15. As n increases, F_n will deviate from Equation 7.12 reaching a plateau. From this, the radius of decorrelation (RDC) can be defined as the intercept of F_{15} and the $F_{n,t}$ line. Figure 7.1 shows fluctuation (SNR is plotted) measurements with an RDC of 7.8 pixels. A larger RDC is better, indicating the spatial extent of structural noise. Alternatively, $1/F_n$ can be plotted against ROI dimension N to give a linear plot.

Friedman and Glover (2006) devised an fMRI QC protocol for a multicentre collaborative fMRI research project, the Biomedical Informatics Research Network (BIRN), that includes measurements of SNR, signal-to-fluctuation noise ratio (SFNR), percent fluctuation and drift, and Fourier analysis of noise residuals plus the radius of decorrelation (RDC) based on Weisskoff's method. The BIRN software tools for fMRI QC are available from http://www.nitrc.org/projects/bxh_xcede_tools [accessed 2 Nov 2014].

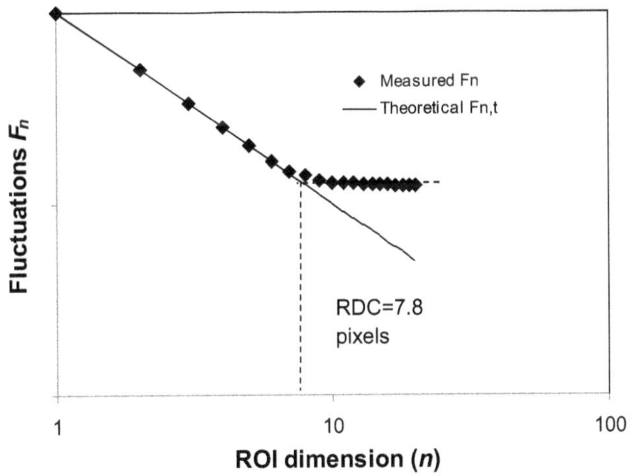

Figure 7.1 Signal fluctuations (F_n) plotted against n (ROI size). The RDC (radius of decorrelation) is the intercept of the plateau portion with the initial slope

7.4.3 Procedure

The Function Biomedical Informatics Research Network (FBIRN) methodology (Friedman and Glover, 2006) specifies a 17-cm uniform agar gel-filled spherical phantom doped with $NiCl_2$. In practice, any uniform phantom could be used although care would be required for a fluid phantom to avoid movement of the liquid.

An fMRI time series is acquired consisting of 200 × 28 slice volumes, 4-mm slice thickness with a gap of 1 mm. The image matrix is specified as 64 × 64 for a 220-mm FOV. TR is 2000 ms and TE is 30 ms for scanners of 3 T and above, or 40 ms for 1.5 T scanners.

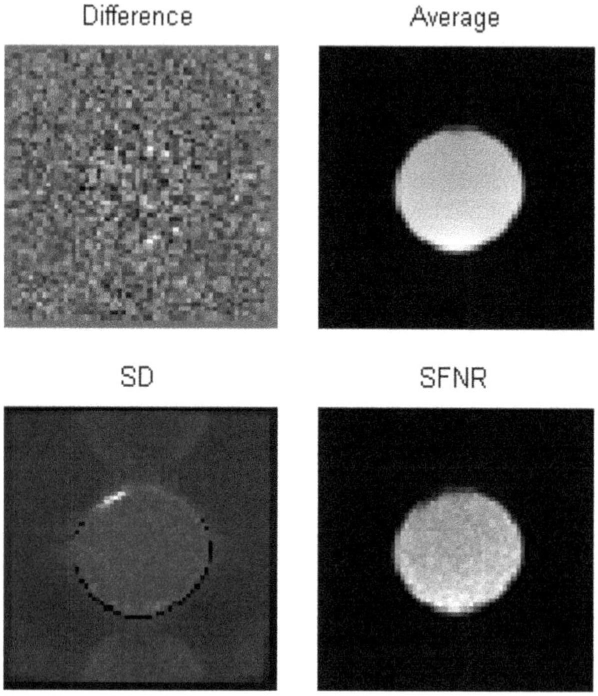

Figure 7.2 FBIRN summary images. The Difference image is calculated from Equation 7.13. The Average and SD images, respectively, represent the mean and standard deviation of pixels across the time series. The SFNR image is given by the Average image divided by the SD image

The FBIRN procedure reduces the time series data into single parametric images of the mean (averaged over the time series), standard deviation, noise average and SFNR (Figure 7.2). The static spatial noise (or difference) image is produced by subtraction of the even images from the odd images.

$$I_{noise} = \frac{1}{N}\sum_{n=1}^{N/2}(I_{2n} - I_{2n-1}) \tag{7.13}$$

The SFNR image is calculated as the mean image divided by the standard deviation image (after detrending). These are shown in Figure 7.2.

The time series is interrogated using a 20 × 20 square ROI centred within the middle slice of each volume. The fluctuation is calculated following a linear trend correction based upon the drift. The frequency spectrum of the fluctuations is also produced, as is calculation of the RDC (Figure 7.3). Transmitter and receiver gain and central frequency also form part of the FBIRN QC report. The RDC is calculated as

$$RDC = \frac{F_1}{F_{15}} \tag{7.14}$$

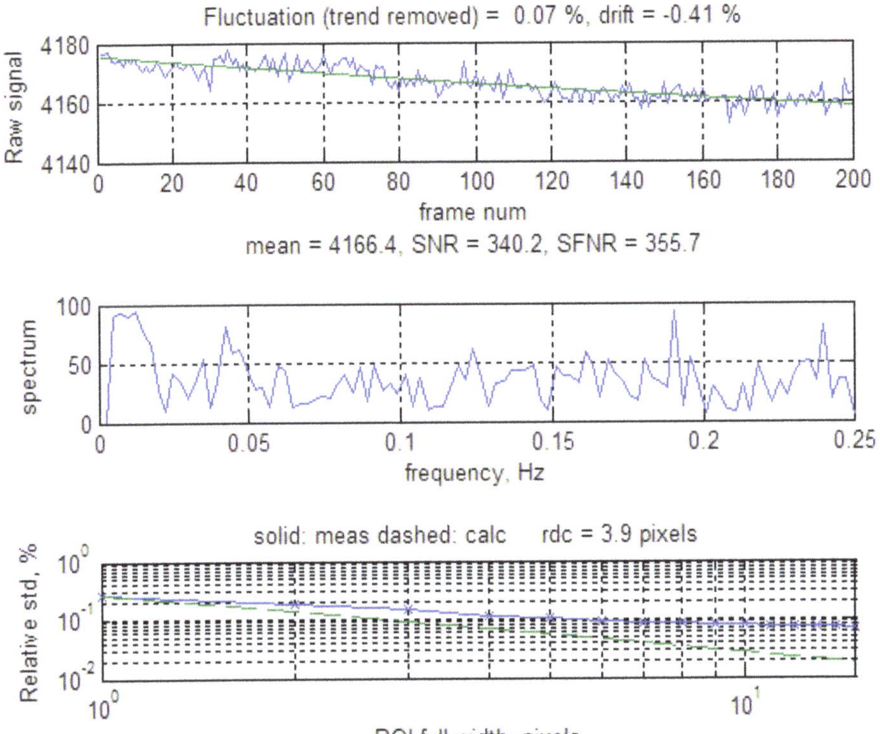

Figure 7.3 fMRI stability QC data. Top: the fluctuation signal over a 200 image time series. Middle: the frequency spectrum of the fluctuations. Bottom: the deviation of the fluctuations F_n (points and blue line) from the static SNR, $F_{n,t}$ (green line) showing a radius of decorrelation (RDC) of 3.9 pixels

7.4.4 Human data

As there are many non-instrumental factors which affect the outcome in fMRI (see Table 7.1), it is important to apply some QC measures to human datasets, particularly if these relate to individual clinical patients. It is important to review the data visually, possibly in a cine loop, to exclude the presence of significant artefacts.

Most fMRI processing involves spatial realignment, and software packages can provide the realignment parameters in terms of three displacement and three angular terms (Figure 7.4). These should be inspected to check for any motion correlated with the paradigm or for excessive motion. Involuntary motion, e.g. from respiration, can also be detected and, in some situations modelled. This is particularly relevant for resting-state fMRI. Bodurka *et al.* (2007) and Greve *et al.* (2011) have addressed means for assessing the physiological component to noise and scanner instability.

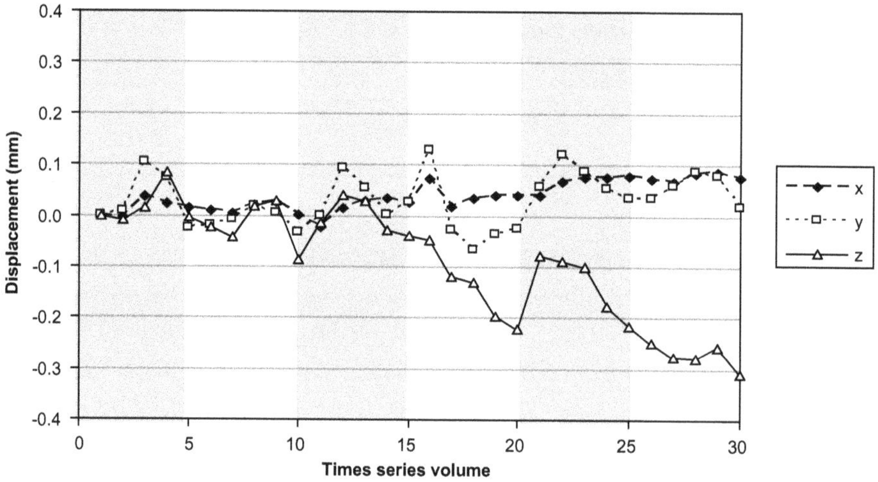

Figure 7.4 Linear realignment parameters for an fMRI acquisition. The paradigm timing is shown by the shading

7.5 Voxel-based morphometry

7.5.1 Introduction

Voxel-based morphometry (VBM) is a technique used to segment MR images of the human brain into different tissue types. The technique is based on the identification of tissue types by voxel intensity classification and is gaining popularity as a parsimonious method by which to investigate differences in neuroanatomy due to ageing, pathology or as an endophenotype in genetic studies. Interpretation of VBM results remains controversial, however, since any changes detected could be a result of phenomena other than those under investigation (cortical thinning or thickening) such as tissue mis-classification, mis-registration or different gyral folding between individuals. The software packages currently available for this type of image processing attempt to address these issues in various ways.

In its simplest form, VBM is based on simple thresholding of images at multiple levels of voxel intensity. More sophisticated techniques use clustering algorithms in conjunction with probabilistic models based on information from both histograms of voxel intensity and neuroanatomical information (Ashburner and Friston, 2000; Smith *et al.*, 2004) (Figure 7.5).

All of the techniques, however, are sensitive to a lesser or greater extent, to partial volume effects, distortion and artefacts in the images and can be partly corrected using post-processing techniques available within the VBM pipeline.

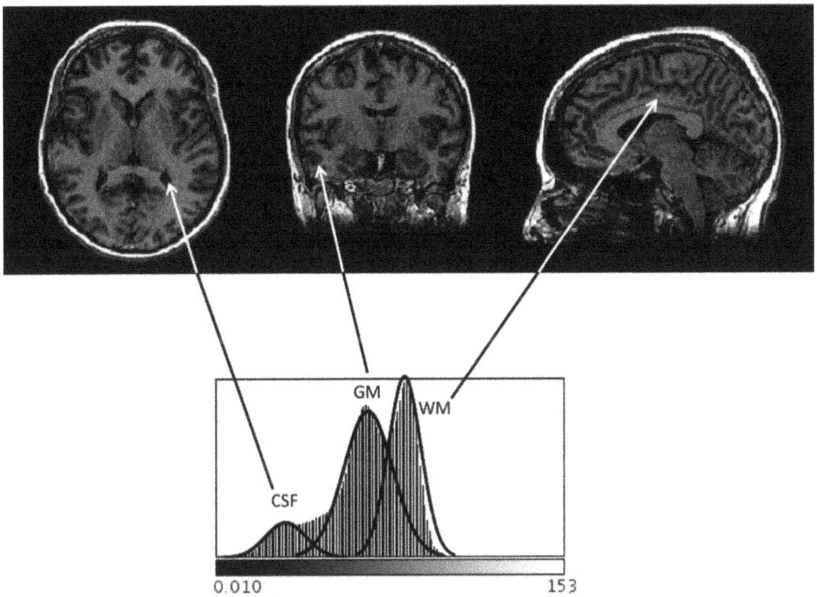

Figure 7.5 (top row) 3D section through a T_1 weighted spoiled gradient echo image acquired on a 1.5 T GE Signa system, 8 channel head coil, ASSET-on, NEX=0.5, TR=35 ms, TE=5 ms, Flip Angle =35°; slice thickness 1.5 mm and in-plane voxel size 0.94 × 0.94 mm and (bottom row) the corresponding histogram of voxel intensities. Tissue segmentation is achieved by estimating the tissue class means and standard deviations using mixture modelling of Gaussians and K-means clustering algorithms. Additional information can be supplied by spatial neighbourhood information of surrounding voxel intensities and the use of standardised tissue probability maps created from averaging many hundreds of previously segmented data. CSF, cerebrospinal fluid; GM, grey matter; WM, white matter

The latest VBM processing pipelines of MR images of the brain will consist of the following basic algorithm:

1. Classification of tissue classes: grey matter, white matter, cerebrospinal fluid, and other (optional), to produce tissue probability maps from high resolution T_1-weighted and/or T_2-weighted images of the brain.
2. Non-linear transformation of each tissue probability map into a standard anatomical space via matching to a tissue-specific average template.
3. Creation of a study-specific template for each tissue probability map.
4. Nonlinear transformation of subject's original tissue probability map to standard anatomical space via matching to the study specific tissue class template.
5. Correction for local contraction and expansion of anatomy by decreasing or increasing voxel intensity correspondingly to produce tissue volume maps.
6. Smooth transformed tissue maps using smoothing kernel of FWHM between 1 and 4 voxels, to account for slight mis-registration and to conform to data distribution assumptions when using parametric statistics.
7. Perform parametric or non-parametric statistics at each voxel in the image.

There are slight variations to this algorithm depending on which software package is used. For example, FSL (http://www.fmrib.ox.ac.uk/fsl/) removes any non-brain tissue before the tissue classification stage whereas DARTEL (http://www.fil.ion.ucl.ac.uk/spm/) leaves non-brain tissue in the image and uses an iterative technique to produce the best study-specific tissue class templates.

7.5.2 Factors affecting grey matter volume measurements obtained using VBM processing

As well as image distortion and artefacts, the success of VBM methods is dependent on the signal-to-noise (SNR) and contrast-to-noise ratios (CNR) and uniformity of the images to be processed, which are in turn dependent on hardware choices such as type of coil used and acquisition parameters such as pulse sequence type and how *k*-space is filled. The change in these image properties due to differences in acquisition parameters can have a profound effect on the performance of VBM algorithms. Figure 7.6 shows the effect of four different T_1-weighted gradient echo (GE Signa system operating at 1.5 T) acquisition protocols on the skull-stripping step of a particular VBM processing pipeline. The error evident in the bottom row of

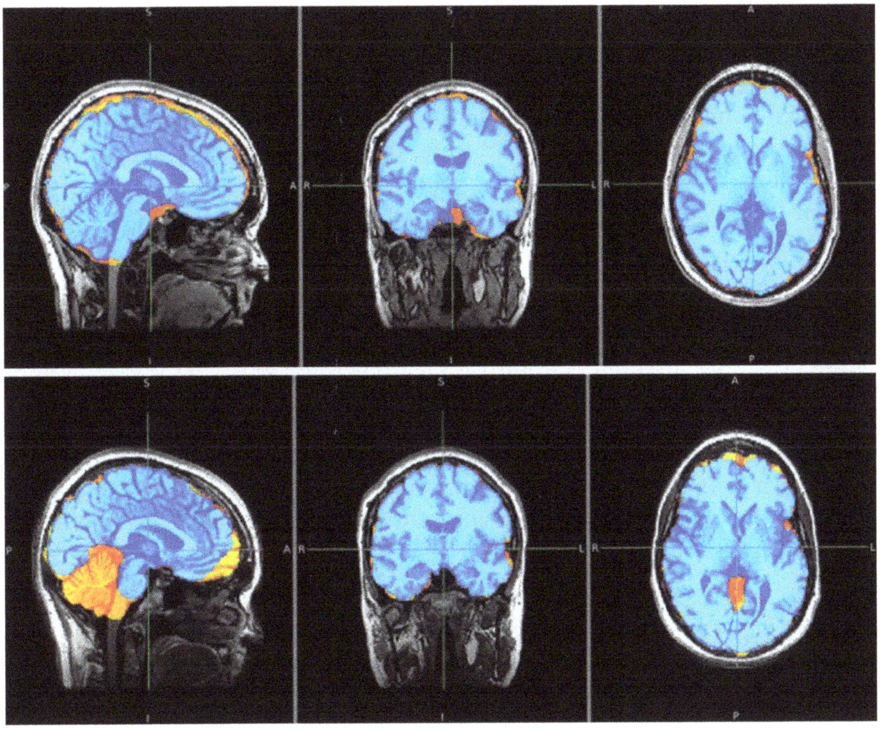

Figure 7.6 Results from using the same VBM pipeline to strip the skull from four different fast SPGR T_1-weighted images of the brain for a single subject (1.5 T GE Signa). In the top row, the skull stripping step from the acquisition performed with a single channel head coil is shown in yellow and the second from the acquisition performed with a multi-channel head coil is overlaid in blue. In the second row, the skull stripping step from the acquisition performed with a multi-channel head coil and NEX=1 is shown in yellow and the second from the acquisition performed with NEX=0.5 is overlaid in blue

the figure can be drastically reduced by altering the post-processing parameters for voxel intensity boundaries or by using an alternative processing pipeline that excludes the skull stripping step. The subtle difference in the results evident in the top row, due to moving from a single channel head coil to a multi-channel head coil, however, is more difficult to correct for. The consequence of this type of error is particularly relevant to longitudinal studies of the brain where a change in hardware may have occurred during the two different time points. In such a case, any subsequent measurements of tissue volume would necessarily include changes due to both image quality and biology/pathology.

Although there are no specific phantoms for assessing and/or optimising VBM protocols, MRI scanners involved in VBM acquisition for clinical trials or research studies should ensure that an initial QC check of their system should include SNR (see Chapter 3), uniformity (see Chapter 4), spatial resolution (see Chapter 4) and slice profile and width (see Chapter 5) using the desired sequence and appropriate phantom. Regular checks should be made during the study to monitor the equipment for any general deterioration in performance over time (Friedman and Glover, 2006).

It should also be noted that performance of these algorithms can vary between subjects for the same acquisition and processing pipelines, and therefore a "calibration study" can be a useful surrogate (Suckling *et al.*, 2010; Moorhead *et al.*, 2009; Stonnington *et al.*, 2008). Data from several normal healthy controls should be collected before embarking on a large-scale study so that approximate performance measures can be obtained and acquisition protocols adjusted where necessary. For example, various scan time-saving strategies can be used (e.g. partial Fourier, parallel imaging, etc.) but the performance of a chosen pipeline may be seriously compromised.

7.6 DCE-MRI

7.6.1 Introduction

Dynamic contrast-enhanced (DCE-) MRI involves the injection of a paramagnetic contrast agent whilst sequential T_1-weighted images are acquired (Sourbron and Buckley, 2013; Jackson *et al.*, 2005; Larsson *et al.*, 2001). The change in signal which occurs can be converted to contrast agent concentration using estimated pre-contrast T_1 values which can be obtained using a number of different methods (Larsson *et al.*, 2001; Brookes *et al.*, 1999; Shin *et al.*, 2009). The accuracy of the pre-contrast T_1 values obtained as well as the temporal stability of the dynamic series can affect the accuracy of parameters obtained from a DCE-MRI study.

7.6.2 Method

A suitable contrast phantom – such as Eurospin phantom TO5 – can be used to assess the accuracy of T_1 values. The phantom should be positioned and imaged using the T_1 mapping sequences as a patient in the DCE-MRI study would be. T_1 values are then calculated and compared to the known phantom values to assess the accuracy of the measurement sequence. Large deviations from the known values should be investigated. The temperature of the scan room should be measured before the investigation as this is known to affect the T_1 values of the phantom. The same phantom can be used to assess the stability of the dynamic time series, which should be run in its entirety. A measure of the signal variation over the dynamic time series can be obtained. Large deviations in signal should be investigated. QC should be run before starting a DCE-MRI study, then on a regular basis during the study, as well as following any scanner upgrades.

7.6.3 Assessment

The ultimate quality control assessment of a DCE-MRI procedure is an assessment of its reproducibility when repeated in the same patient under the same conditions. This is particularly important when assessing changes in DCE-MRI parameters following treatment. Although logistically difficult, consideration should be given to assessing the reproducibility of a protocol in a small subset of the patient population early on in a study (Galbraith *et al.*, 2002; Jackson *et al.*, 2003). Scans should be performed as close together in time as possible and without any treatment having been administered in between. Coefficients of variation in the parameters obtained should be minimal and can be compared to those obtained in Galbraith *et al.* (2002) and Jackson *et al.* (2003).

In recent years, there has been some work in the development of flow phantoms for assessing the accuracy of DCE-MRI (Rajan *et al.*, 2014; Hariharan *et al.*, 2013; Freed *et al.*, 2011). These phantoms aim to replicate flow rates in vessels and/or tissues so that an assessment of both the scanning protocol and analysis methods for arterial input function and subsequent tracer kinetic parameter calculation can be made. Care should be taken that the flow phantom used replicates as closely as possible the tissue/vessel conditions under investigation.

7.7 Phase-contrast velocity mapping

7.7.1 Introduction

Phase-contrast velocity mapping or flow imaging is a phase-difference method, based on repeated imaging with modified velocity sensitivity in the relevant direction (Biglands *et al.*, 2012). The modification is typically a bipolar pair of opposed amplitude gradient pulse areas giving a phase shift proportional to the change in position of excited magnetisation, but nominally zero phase shift if static. The repeat is usually time-interleaved to minimise other differences between the images being subtracted. The method is generally cardiac gated for cine velocity mapping of blood, CSF or myocardium, but may be ungated where there is lower pulsatility. The velocity sensitivity (the velocity causing a phase shift of 180°, known as VENC) must be adapted before scanning: in CSF work, VENC can be ~5 cm/s, whereas an aortic stenosis can require VENC 600 cm/s. A smaller VENC requires a greater gradient waveform difference between the images. As with much of MRI, some aspects of phase-contrast imaging are incompletely tested for official safety certification.

7.7.2 Methods

The manufacturer should have evidence that the initial sequence design was tested in the relevant software revision level and type of scanner. Velocity sensitivity is based on gradient waveforms that are monitored by routine maintenance checks of gradient calibration at each site, and no further checks are necessary if that initial proof of correct velocity calibration is clearly documented by the manufacturer as expected. However, velocity-encoding modifications by users should be tested independently of the gradient waveform design. This separate check might be by simulation using the pulse sequence waveforms, or by setting up a flow phantom, typically timing a measured volume of collected fluid as "gold standard", known commonly as the "bucket and stopwatch technique".

Such flow phantoms are marketed at substantial expense but can easily be "home-brewed" for steady flows. Apart from long tubes to keep the pump out of danger, the fluid should be in the B_0-field for sufficient time to be magnetised before reaching the image slice. For velocity calibration from flow, the image FOV must be correct. It is unnecessary to match blood viscosity or relaxation times, since velocity distribution across the tube and absolute SNR are not the aim. Highly pulsatile flows as in the main arteries are much more difficult to model with phantoms, as the distensibility of long tubes from isocentre to the pump outside the MRI room tends to dampen pulses with impracticably extreme pre-emphasis required to correct this. Temporal resolution has been studied in cine flow imaging for sufficiently accurate capture of the systolic flow peak (Lotz *et al.*, 2002). The construction of realistic stenotic jet phantoms demands careful attention to phantom replication of the shape of the stenosis suited to the disease in question (e.g. valve defects, coarctations).

Most phase-contrast work uses spoiled low flip-angle gradient-echo sequences with short TE, so the technique is robust with respect to scanner adjustments such as shimming, reference-frequency, B_1, and receiver coil profiles. Pitfalls in flow analysis software are common, for example, ROI inclusion of border pixels corrupted by air noise or fat signals, and any changes in the flow analysis software used by medics should be assessed.

7.7.3 Background errors

This is perhaps the only topic where scanner QC at some centres might still be necessary, depending greatly on the platform and protocol used (for example, a switch from free-breathing to breath-hold imaging could require reassessment).

The phase difference due to velocity is contaminated by three systematic sources of phase shifts (in addition to random noise). First, concomitant gradient terms, also known as Maxwell terms, are distortions of B_0 occurring with the MRI gradients. Their impact in phase-contrast can be solved analytically (Bernstein *et al.*, 1998) near isocentre assuming linear gradient fields; this solution is usually included by vendors. This correction should be straightforward, but it is difficult to ensure that all circumstances are free from bugs. An exceptional arrangement with zero Maxwell-term difference uses 'symmetric' velocity-encoding waveforms that furthermore must not overlap other gradient pulses (Bernstein *et al.*, 1998).

The second source of phase errors arises from incomplete spoiling of transverse magnetisation between the rapid iterations of the gradient-echo sequence. It is not a distortion of B_0 but is a weak refocused transverse magnetisation underlying the direct gradient echo. The weak signal has a highly complex phase behaviour affecting the total measured phase but it seems negligible in most cardiovascular work (Gatehouse *et al.*, 2012).

The third source of error, usually cited as "eddy currents", is due to small distortions of B_0 remaining after the techniques of active gradient-coil shielding and pre-emphasis have vastly reduced the initial problem or slightly over-corrected it. This error has a complex spatial and temporal-decay on timescales of many seconds to sub-millisecond, or even oscillatory behaviour (Bernstein *et al.*, 2004) or correct an incorrect citation.). The pre-emphasis calibration is understandably approximate and has many degrees of freedom which are often overestimated. While this calibration is not expected to change greatly with time, some short-term impact of heating by eddy currents in prolonged scanning can perhaps not be ruled out.

Curvature in the background error across the FOV may originate from any of the three sources described above. Maxwell-term errors (type 1) should not change with cine frame or sequence repeat time, provided that the slice location and orientation are held constant and no waveforms are changed during the velocity-encoding. The second and third sources of error can change between cine frames or with repeat time; for these, retrospective gating would be expected to show reduced inter-frame variability compared to a truly pausing prospectively triggered cine.

On clinical systems, the background error may easily reach $\pm$ several percent of VENC. If the velocity imaging was acquired with VENC reasonably well-matched to the peak velocity of interest, the background error is negligible for peak velocity assessment, with obviously greater impact on lower velocities. However, for volume flow measurement in a large vessel over a long cardiac cycle, the integrations over regions and times of lower velocities can magnify the error so that even 1 cm/s background error can disrupt flow by 10%. As these errors tend to reduce nearer isocentre, systems usually move the patient table for phase-contrast sequences so that the centre of the image slice is in the $z=0$ plane. (Ideally, the *vessel of interest* in the image should be moved to the $z=0$ plane.) Measuring flow in vessels highly offset along x or y may require extra attention to patient positioning, as with large offsets there is a further impact of gradient non-linearity on velocity-encoding accuracy (Markl *et al.*, 2003). Another approach often cited as reducing the problem is to slow down the pulse sequence, so that its lowered gradient amplitudes excite smaller eddy currents. However, the background phase error arises from the time integration of the eddy currents, leading to a complex dependence on the time-constants of the eddy currents in relation to the duration of the velocity-encoding pulses and also the time after them to the gradient-echo.

If there is enough stationary tissue in the image, it enables fitting of a background correction across the FOV which is subtracted to correct the velocities (Lankhaar *et al.*, 2005). The stationary tissue may be identified automatically or by placing markers. The stationary tissue should be distributed over both directions of the FOV, avoiding phase-encode wraparound, and use regions of adequate SNR. Lung voids in cardiac MRI usually prevent the use of any tissue near the vessel of interest. The fitting accuracy becomes critical for cardiac MRI where this method is often ineffective (Rigsby *et al.*, 2013). A similar filtering approach without requiring users to identify stationary tissue is installed on some systems, and users would be wise to test its unpublished limitations. A "field camera" method applies a distributed set of high-SNR marker probes to measure the error (Giese *et al.*, 2012) but again would depend on spatial fitting to estimate errors deep within the patient, and more work is needed for its routine viability. The prospects for reliable *in vivo* correction remain poor.

The background error is usually ignored in clinical work as it is only one of many factors degrading accuracy. There is large variability of background error between scanners of nominally the same type and sensitivity to imaging parameters including slice location and orientation (Gatehouse *et al.*, 2010; Rolf *et al.*, 2011). This does explain why no upper limit can be expected from manufacturers.

7.7.4 Practical hints for background assessment

If there is concern over background errors in a patient, and the *in-vivo* correction is uncertain, correction can be obtained later by replicating each velocity scan precisely, obtaining its background error in a stationary phantom (Holland *et al.*, 2010; Miller *et al.*, 2009). However, the results are robust with respect to shimming, reference frequency, B_1 and coil sensitivity profiles and there is no requirement to match these to the *in vivo* conditions. The approach of "correcting it only if suspicious" seems acceptable to clinicians who usually know much other data about each patient, but this approach has potential to be misleading. Phantom correction is straightforward and takes a matter of minutes. Naturally, there are contradictory reports about whether it is justified in the clinical environment for every scan, because the error varies with scanner type, routine service recalibrations (e.g. of pre-emphasis), software revision, sequence parameters, image slice orientation, etc. Without rigorous QC, some doubt will always remain over the background error. Temporal instability of the scanner (Gatehouse *et al.*, 2012) reached marginally significant levels over periods tested up to 8 weeks, but corrections would usually be acquired before clinical reporting, with marginally significant difficulty over the short-term thermal-related drift in background velocity errors on certain systems studied further by Busch *et al.* (2014).

For cardiac MRI, a 25-cm diameter spherical phantom is suitable. A short T_1 enables higher flip angle for improved SNR, while otherwise replicating the patient's scan *exactly*. Note that if any automated background correction algorithms are enabled, these are likely to perform differently in phantoms than in patients, *completely invalidating* this method for post-correction or for QC. Large ROIs can be used if necessary, as the error varies only slowly with position, but it can be helpful to copy the clinical vessel ROIs to the phantom so that flow correction can be measured by the same software as used by clinicians, in which case use an ECG simulator at the same rate as the patient. While the velocity error does not depend on heart rate, the error in volume flow does.

Variations in the background error with cine frame should be small unless prospectively triggered imaging is used, where large variations in the first frames may not be simply eddy-current-dependent and some care is needed. Unless the impact of incomplete spoiling is abnormally prominent, there is no need to match *in vivo* relaxation times, B_1 and B_0 fields. The phantom may contain fluid if patient table shifts are followed by adequate delays; 5 s was ample settling time in one example tested, but this should be checked in each set-up. For certainty, an airtight container of doped (by collected residues of Gd-based contrast agent) and disinfected (propylparaben or sodium azide) domestic gelatin at approximately twice its normal jelly-pudding strength has been used (Gatehouse *et al.*, 2010). Do not boil mixtures containing contrast agent or disinfectants; add these only when lukewarm.

Ideally, each scanner would be supported by routine QC of the background error, which might collect a range of slice orientations using a set of protocols that are applied clinically, repeating it only after service events that may affect background errors, such as pre-emphasis recalibration. Unfortunately, some temporal instability of the background errors renders a "database" of various angles and parameters unlikely to be successful (Gatehouse *et al.*, 2012) and it has not been attempted.

7.8 MR spectroscopy

7.8.1 Introduction

In MR spectroscopy (MRS), the quality of the acquired spectrum is very sensitive to any variability in all parts of the MR system and hence regular QC is required to monitor performance so that, where possible, corrective action can be taken in order to give consistent and reliable interpretation of MRS examinations (Bovee *et al.*, 1995; Drost *et al.*, 2002). Such data are also useful for comparing MRS exams acquired on different platforms (De Stefano *et al.*, 2007; Van der Graaf *et al.*, 2008).

7.8.2 Phantom

The manufacturer-supplied QC phantom typically consists of a spherical container of a solution of a number of MR-visible chemicals of known concentration, at least one of which is weakly J-coupled. The spectral parameters that should be routinely monitored are SNR, level of water suppression, line width and shape, relative concentration and, if such an approach is being used, absolute concentration, and chemical shift.

7.8.3 Methods

The most straightforward method is to use the same pre-scan adjustments, acquisition protocol and post-processing that are used *in vivo*, and an acquisition without water suppression should also be used. The shortest TE available will give the most sensitive test. Line width (usually full width at half maximum in Hertz) is a sensitive measure of both field homogeneity and the presence of eddy currents and consideration should be given to localising both at isocentre and a reproducible off-centre location. Line shape, especially asymmetry reflects the presence of eddy currents and a simple measure of this is the "goodness-of-fit" parameter in the line-fitting software used. This assumes that, if available, no eddy current correction is applied – see below. How the line shape of any J-coupled moiety varies with TE will depend on the specific implementation of the sequence. When assessing chemical shifts, it should be remembered that the test phantom is at room temperature rather than 37℃. Peak area measurements relative to each other and to unsuppressed water should be consistent and linear. For the SNR measurement, the ratio of peak area can be compared to root mean square noise in any suitable baseline region of the spectrum. For longitudinal comparisons and for comparisons between different scanners, consistent filtering, as far as possible should be used.

More sensitive measures can be achieved by analysing data with the minimum of post-processing filtering, ideally just zero-filling and phasing. For instance, a more meaningful measure of level of water suppression is achieved by comparing the unsuppressed and suppressed water peak without the additional filtering applied to this part of the spectrum. However, it should be remembered that much longer T_2* values will likely be obtained in the phantom than *in vivo* and this may need to be reflected in the acquisition window.

Spatial localisation in PRESS, STEAM and, with the addition of phase-encoding gradients, chemical shift imaging sequences depends on the combination of shaped RF pulses and applied gradients. Clearly, the localisation achieved in either single- or multi-voxel spectroscopy is an important quality parameter but suitable test objects are not widely available. Some solutions can be found in the literature (Leach *et al.*, 1995; Hunjan *et al.*, 2003; Woo *et al.*, 2007) using various forms of compartmentalised solutions.

Only proton spectroscopy has been addressed here but where MRS of other nuclei such as ^{31}P or ^{13}C is being used then, with the exception of references to water suppression, similar considerations apply.

Finally, it is worth noting that the MR system is in essence a spectrometer and, as such, simple spectroscopic procedures can give a very sensitive overview of overall system performance. For instance, variation in the residual water peak from excitation to excitation is a very sensitive indicator of any RF instability, whilst drifts in the position of the unsuppressed water peak over extended periods of time reflect field variation most likely arising from temperature fluctuations in the magnet hardware.

7.9 QC at ultra-high field

7.9.1 Introduction

The principal reason for moving to higher field is the increased SNR. Magnetic polarisation increases in proportion to the field strength, so 7 T should have more than double the available signal of 3 T. However, this does not always translate to a similar increase in SNR or contrast-to-noise ratio (CNR), and the engineering challenges are significant.

As the field strength increases, tissue T_1 values increase and tend to converge since the molecular tumbling rates are lower relative to the Larmor frequency and relaxation is less efficient. T_2*, and to some extent T_2, values decrease with higher field. Pulse sequence parameters therefore need to be adjusted for operation at high field if image contrast is to be comparable with that at lower fields. Echo times should generally be shorter and repetition times longer. One advantage of the longer T_1 values is increased CNR in time-of-flight angiographic sequences since the background tissue signal is more readily saturated relative to signal from inflowing blood.

The water–fat chemical shift increases (in absolute frequency terms) with field strength, and hence the chemical shift artefact becomes more obvious in images unless the readout bandwidth is increased in proportion.

Susceptibility effects increase strongly with field strength, and so image distortion and related artefacts generally increase unless echo times are reduced to the minimum possible. Shorter readouts can be achieved using parallel imaging techniques. Increased susceptibility effects are of course helpful for techniques that depend on susceptibility to create contrast, notably Blood Oxygenation Level Dependent (BOLD) and susceptibility-weighted imaging (Haacke *et al.*, 2004). The increased BOLD effect in fMRI studies is one of the main drivers behind the interest in 7 T scanners. Although background "physiological" signal fluctuations also increase, BOLD signals can be detected from correspondingly small pixels that are dominated by thermal noise. Thus, the theoretical SNR improvement is best used to increase spatial resolution.

7.9.2 Ultra-high field QC

Many 7 T sites are still at the stage of technical development rather than clinical use, and QC procedures are not yet in routine practice. The main difference relative to lower fields is in the behaviour of the RF (B_1) field. In particular, it is extremely difficult to produce homogeneous excitation from an integrated body coil as used in scanners at 3 T and below. Instead, many 7 T systems use localised coils for transmission as well as reception, and the transmit efficiency and receive sensitivities must both be considered. Because of the inhomogeneous properties, SNR performance cannot easily be characterised by a single number. Not only is signal strength non-uniform, but so is the noise in parallel imaging techniques, and so SNR measurements must be made locally at the position of interest. A useful method for measuring SNR is to compare an image acquired using RF excitation with an image acquired with excitation switched off (Dietrich *et al.*, 2007).

The actual values of SNR achieved are a complex function of field strength, sequence and gradient characteristics, coil design and loading. In a recent comparison of Diffusion Tensor Imaging at 1.5 T, 3 T and 7 T, for example, Polders *et al.* (2011) found the increase in SNR between 3 T and 7 T to be greater than the expected 7/3, and attributed it to the increased quality of the receive coil array on the 7 T scanner.

Scanner software typically includes homogeneity correction filters that may be applied to the images either as part of the acquisition process or retrospectively. Care must be taken to ensure consistency when measuring QC parameters: have the filters been used or is this a "raw" uncorrected image? (see Section 3.3.1). Increasingly, parallel-transmit technology (see Section 3.3.1) is being used to minimise the effects of non-uniform RF excitation. Comprehensive QC of such systems requires separate measurements for each channel.

QC procedures for fMRI are designed to assess image stability over the time-course of a stimulus-activation paradigm of typically 10 min. BOLD effects of a few percent must be detected, and so stability of the order of 1% is required. Echo Planar Imaging is used to provide sufficient temporal resolution, and it places heavy demands on the gradients. SNR, ghosting and drift are parameters of interest (Weisskoff, 1996).

7.9.3 Phantom design

At high field, water-based phantoms exhibit significant signal non-uniformity because of dielectric effects (see Section 3.4.2). Tests objects for flip angle and uniformity measurements should therefore be oil filled (see Section 2.3), since oils have much lower permittivity thus minimising standing waves. Silicone oil is preferred since it has a single proton resonance frequency unlike vegetable oils that have multiple resonances that may confound measurements. The amount of any dopant will need to be adjusted if physiologically realistic relaxation times are required. For longitudinal tests of stability, readily available water-filled phantoms may be used and the non-uniformities simply accepted.

The commonly used plastic poly-methyl methacrylate ('Plexiglass', 'Perspex') continues to be useful for construction of homogeneously filled phantoms since its susceptibility (–6 ppm) is not too different from those of oils and water (around –9 ppm). The effects are more of a problem for internal structures when measurements should be avoided close to the plastic–liquid interfaces. Susceptibility effects also depend on the orientation of the structures relative to B_0. Great care is required to eliminate air bubbles since they cause significant artefacts (see Section 2.2.1).

7.9.4 Ultra-high-field spectroscopy

Increased sensitivity and increased spectral dispersion make ultra-high fields attractive for spectroscopy, provided that the technical hurdles can be overcome. Spectral linewidths generally increase with field strength and so the theoretical SNR improvement is not achieved in practice. Spectroscopic multiplets arising from J-coupling (for example, the lactate doublet) collapse and generally cannot be resolved, unlike at lower fields. Large excitation bandwidths are required to excite the coupled protons of J-coupled pairs, and the corresponding power requirement may exceed that available from conventional RF pulses. One solution is to use adiabatic pulses which combine high bandwidth with (relatively) low peak power, as demonstrated recently at 3 T (Scheenen *et al.*, 2008). Any prior knowledge used in spectral quantification should be appropriate for the field strength and echo time used.

For spectroscopy (and imaging) at very short echo times, care must be taken to ensure that the phantom materials (and coil assembly) do not themselves contribute unwanted signal (Marjanska *et al.*, 2008). For example, PTFE (Teflon) material may be used to eliminate the background proton signal, but must be avoided if measuring fluorine compounds using ^{19}F spectroscopy.

Acknowledgements

The authors acknowledge the following in the preparation of this chapter – Daniel Polders (Philips Healthcare, Nijmegen, Netherlands), Oliver Kraff (Erwin Hahn Institute for MRI, University of Duisburg Essen), and Stuart Clare (Oxford Centre for Functional MRI of the Brain, University of Oxford).

Part II:
Artefacts in MRI

8 Artefacts

Geoff Charles-Edwards, Martin Graves and Nick Weir

8.1 Introduction

In general terms, an image artefact in MRI may be considered to be any unwanted image feature that does not represent the true anatomical structure or physiological information on the subject or object being scanned. However, it is important to remember that what might be considered an artefact in one situation can be exploited to provide useful information in another, e.g. blood oxygenation level dependent (BOLD)-MRI and susceptibility-weighted MRI. Due to the complexity of MRI, there are many potential causes and manifestations of artefacts that may present as significant image degradation or perhaps more worryingly as subtle changes that may be mistaken for pathology. Here a range of MRI artefacts are described, grouping them according to their principal source:

- system faults and hardware limitations;
- the patient;
- physics limitations;
- limitations of image reconstruction techniques.

There is not always a clear separation of these groups, and certain artefacts, e.g. RF spikes and geometrical distortions can be caused by more than one of these categories. As a user of MR equipment, it is important to be able recognise and understand these artefacts, and to know the various options to manage them even though it may not always be possible to remove them completely. The choice of pulse sequence and user-defined parameters can have a significant effect on whether an artefact is exaggerated or minimised. In many cases, a good grasp of the relationship between *k*-space and image-space is very helpful to understand the cause of an artefact and determine potential solutions, and it is worth noting that modern MRI systems typically employ several artefact reduction schemes automatically.

In this chapter, the main artefacts that may be encountered in clinical MR and quality assurance are described, together with human and phantom examples. Where relevant, basic steps that can be taken to eliminate or reduce the artefact are also given, utilising functionality that is currently available on clinical MRI systems. If artefacts originate from a hardware fault, then this will usually come under the remit of the service contract. Fortunately, some equipment-based image artefacts have all but disappeared from modern MRI systems through improved hardware, although these have been replaced with new image artefacts associated with increasingly sophisticated image reconstruction techniques.

8.2 System faults and hardware limitations

8.2.1 Continuous RF interference – 'zipper' artefact

Since the MR signal that we aim to measure is very weak, MRI is inherently very sensitive to RF interference from sources other than the patient. Interference may arise from outside the room if the Faraday cage is compromised, from RF emitting electrical ancillary equipment within the scan room (e.g. faulty patient monitoring, ECG equipment, RF ablation equipment) or from a problem with the scanner itself. The resulting artefact is often called 'zipper' artefact (although the term zipper is sometimes also used to describe other artefacts with similar appearance) as it usually manifests as one or more lines of interference across the image in the phase-encoding direction, as highlighted in Figures 8.1–8.4, although these lines may appear slightly curved in distortion corrected images. The frequency of the interference is characteristic of its source and determines the position of the lines in the frequency-encoding direction.

To avoid external RF getting into the Faraday cage, MR operators should ensure the scan room door is properly closed during the examination and avoid positioning any conducting material through waveguides. Using only ancillary equipment certified for use in a MRI magnet room is essential. If the source of interference cannot be removed from within the Faraday cage, it may be possible to minimise or remove the artefact from the images by reducing the receiver bandwidth and/or adjusting the central frequency of the scanner.

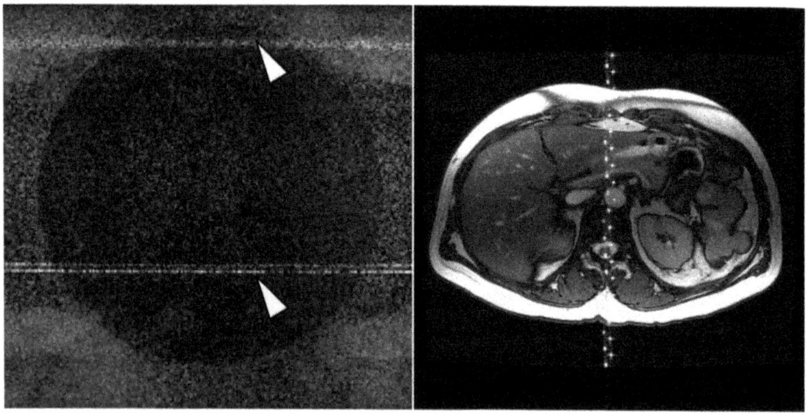

Figure 8.1 RF interference ('zipper') artefact due to a non-MR-compatible flow pump (left panel). A standard quadrature birdcage coil was used with a flow imaging sequence (hence zero signal from stationary phantom). Phase-encoding was in the horizontal direction. RF interference from an adjacent MRI scanner when the magnet room door was open (right panel). Phase-encoding direction was in the vertical direction

Figure 8.2 Image from an MRI sequence with high matrix and maximum bandwidth shows RF interference from a faulty MRI ECG battery pack (left panel) that disappeared when the pack was replaced (centre panel). RF interference at multiple frequencies arising from outside the Faraday cage when the magnet room door was open (right panel) which similarly disappeared once the door was closed. The phase-encoding direction in all cases was vertical

Acquiring an image with a high matrix and maximum receiver bandwidth can be used to probe for RF interferences over the widest range of frequencies. A length of non-ferromagnetic conductor, e.g. a power cable, can be placed on the MR table top to reproduce the antenna effect of a patient in the scanner. To probe low-level RF interference, it is desirable to maximise the receiver gain. However, the automatic preparation stages immediately before the MRI sequence may need an object to be placed in the magnet to provide some signal, so minimal flip angles or signal suppression can be used to minimise this in the imaging sequence.

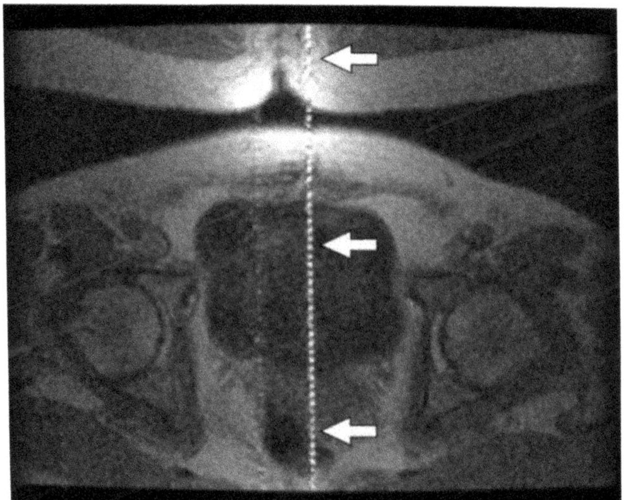

Figure 8.3 RF feed-through, appearing in the centre of the image at the Larmor frequency. Phase-encoding was in the vertical direction

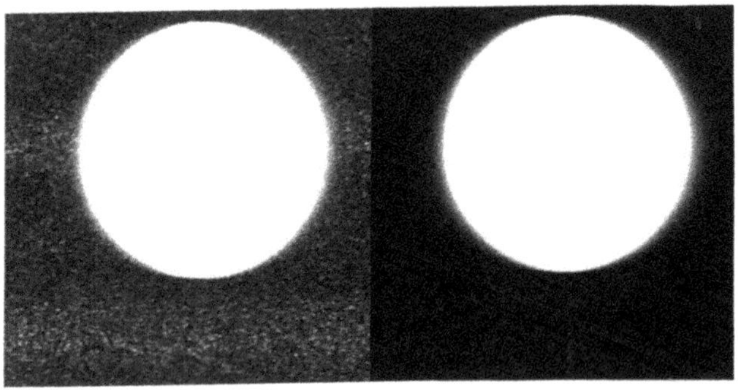

Figure 8.4 Reducing the window level shows low level RF interference following replacement of halogen light bulbs in magnet room with general purpose LED bulbs designed for conventional light fittings (left panel). The interference disappears on a repeated image acquisition when the lights are turned off (right panel). Phase-encoding was in the horizontal direction. Both images have the same window levels. LED lighting certified for use in MRI scan rooms should not produce RF interference

It is becoming increasingly common for MRI units to have more than one scanner. To avoid the high power transmitted RF signals from one MRI scanner appearing as potential RF interference on other scanners, it is prudent to set the centre frequencies of each system to be as far apart as possible within their permitted range. To test for cross-talk between MRI scanners in adjacent rooms, acquisitions using high intensity RF can be performed simultaneously on both scanners (Zhuo and Gullapalli, 2006).

Although associated more with older generations of MRI scanners, a specific example of RF interference that arises from the MRI scanner itself is RF feed-through. Here the signal from the RF amplifier is not adequately gated off during signal acquisition and is detected, appearing at the Larmor frequency at the centre of the image, as shown in Figure 8.3. If observed, it should be possible to remove it via phase cycling, something that may be applied automatically with even numbers of signal averages.

8.2.2 Instantaneous RF interference – RF spike/'corduroy'/'herringbone' artefacts

If the RF interference occurs as a short burst, such that it is only present during the acquisition of a single data point in *k*-space, creating a so-called RF spike, then the resulting image artefact will present very differently to that described in Section 8.2.1. Since each point in *k*-space describes the contribution of a particular spatial frequency to the whole of the MR image, RF spikes result in the formation of regular stripes across the whole image, as demonstrated in Figure 8.5, although additional signal processing widely employed in clinical MRI systems may vary the intensity of the artefact across the image. For an individual RF spike, the distance and direction from the centre of *k*-space will determine the frequency and direction,

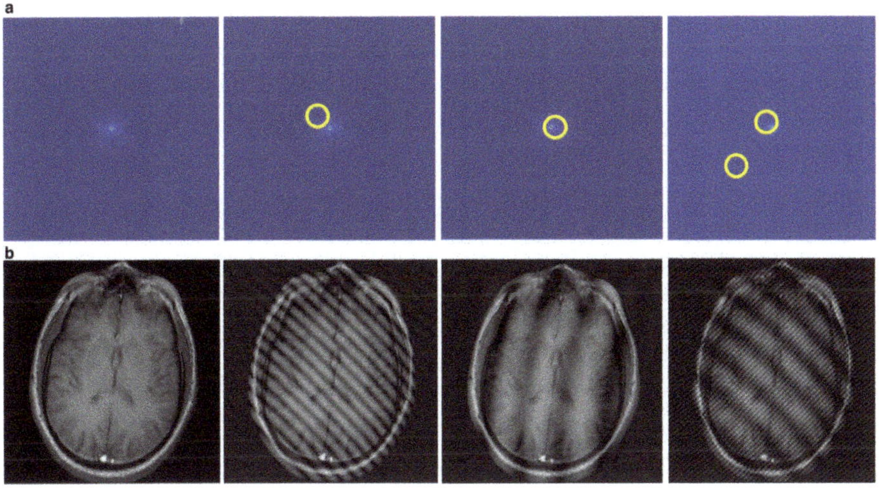

Figure 8.5 (a) k-space for an axial brain slice, (b) equivalent post FFT image, with different simulated RF spike scenarios, in order from left to right: clean, single RF spike, single RF spike closer towards centre of k-space, two RF spikes. The position of each spike in k-space, indicated by the circles, gives rise to regular stripes across the MR image also known as a corduroy artefact. Spikes close to the centre of image result in very low spatial frequency signal variations across the image. Multiple RF spikes will result in the superposition of the corresponding spatial frequencies of each spike onto the image, giving rise to a so-called herringbone artefact (bottom right panel)

respectively, of the stripes. Such artefacts are sometimes known as a corduroy artefact. Multiple RF spikes during the same image acquisition will result in a combination of image stripes, sometimes known as a herringbone artefact.

RF spikes typically arise from electrical discharges that come about from the build-up of static electricity or from faulty electrical components. There are numerous potential causes of RF spikes including faulty lighting/blown bulbs and loose connections in the MRI scanner, e.g. nuts and bolts that hold in place the shim trays or electrical cables supplying the gradient coil, which come under significant stresses during MRI, particularly with gradient-intensive sequences such as echo planar imaging (EPI). Indeed, EPI sequences may be used to search for RF spikes, with the data viewed as a cine loop to quickly search for spike issues. RF spikes may also be caused by static discharges from clothing, or conductive debris in RF coil plugs or faulty components within the coil. This wide range of potential sources of RF spikes together with their intermittent nature can make it difficult to identify the cause of RF spike artefacts.

Faulty or blown light bulbs that may produce RF spikes should be replaced. In some circumstances, the lighting circuitry itself may be faulty and this may need to be completely isolated at the mains distribution unit to diagnose the problem. Many new MRI installations are installing LED room lighting to avoid the need to change bulbs but also to minimise potential RF interference. Ideally, such LED lighting should be installed with the AC to DC conversion outside of the Faraday cage to minimise the risk of more continuous RF noise generation, demonstrated earlier in Figure 8.4.

Air humidity has an important influence on RF spikes, since drier air has a higher dielectric value that allows a larger static charge to build up before an electrical discharge (arcing) occurs, resulting in a larger RF spike and more obvious artefact. Consequently, some MRI manufacturers specify minimum requirements for magnet room humidity, and air conditioning to magnet rooms should typically be fitted with humidifiers. MRI service providers may decline to troubleshoot artefact problems for systems where the humidity is too low.

If the raw data are collected, then there is scope to recover artefact-free images by removing the spikes in the *k*-space data before reconstruction. However, at the time of writing, we are not aware of simple tools to do this from the main MRI manufacturers, and as such, this is not a practical solution for many.

8.2.3 DC offset/baseline errors/spurious signal

A DC offset can be defined as any non-noise signal that is recorded by the MR system that does not originate from the desired MR signal. Often known as spurious signal, it gives rise to baseline errors in the raw data. Such signal may be constant throughout *k*-space, vary in the phase- or frequency-encoding directions, or both. Each of these gives rise to a different artefact in the image.

In its simplest form, a DC offset is a constant addition to the recorded signal everywhere in *k*-space. Since this essentially has zero frequency, it appears only as a single bright pixel in the centre of the image after Fourier transformation. Traditionally, such a DC offset could arise from a small non-zero output voltage bias in the receiver electronics. If this DC offset drifts gradually over time, such that it varies over phase-encoding steps in conventional rectilinear *k*-space filling but is essentially constant over each frequency-encoding readout period, then the central bright dot in the image will spread out in the phase-encoding direction. Fortunately, this problem has essentially been eliminated in most modern MRI scanners by digitising the signal directly or at intermediate frequencies before fully demodulating via digital processing.

DC offsets can also arise from the pickup of signal during the readout period from unwanted signal coherences, most notably from unsuppressed transverse magnetisation freshly excited by non-perfect refocusing pulses. Unlike the constant DC offset, this unwanted signal varies during the readout period, due to both T_2* decay and frequency-encoding. If there are no differences in this signal between phase-encoding *k*-space lines, i.e. the signal is not phase-encoded, then it will appear as a single interference line in the centre of the image spreading along the frequency-encoding direction. Fortunately, this DC artefact can be addressed by phase cycling the intended excitation RF pulse and receiver by 180° on alternating lines of *k*-space. This does not affect the MR signal, but modulates the DC offset at half the Nyquist frequency such that, following Fourier transformation, it appears at the edge of the image, away from areas of interest where it can be hidden by blanking a few lines at the edge of the image, as shown in Figure 8.6. Additionally, phase cycling with paired alternating RF pulse and receiver phases can be used for each *k*-space line to cancel out the DC offset if an even number of signal averages is acquired (Stadler *et al.*, 2007). Such phase cycling may be applied automatically by MRI systems when selecting an even number of signal averages.

Finally, if these unwanted signals are additionally phase-encoded, they can appear over localised regions of the image. Since these unwanted signals are typically unsuppressed fresh transverse magnetisation, they are most intense at the edge of *k*-space. Consequently, as with peripheral *k*-space RF spikes, the resulting artefact is characterised by high frequency variations in the image, and is often known as a fine-line artefact, as demonstrated in Figure 8.7. However, unlike RF spikes, fine-line artefacts will typically cover only a region of the image since the intensity of unwanted signal will be dependent on the relaxation times of the underlying tissue.

Fine-line artefacts are most commonly observed with fast/turbo spin-echo sequences and can also arise from spurious signal as a result of incomplete gradient spoiling of an RF spatial or frequency-selective saturation pulse (Bernstein *et al.*, 2006). Unless spoiling gradients are automatically increased, the larger gradient amplitudes

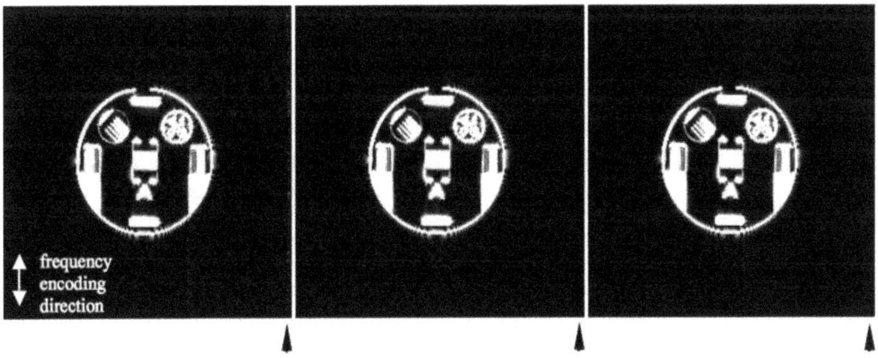

Figure 8.6 DC offset artefact, arising from frequency-encoded unwanted signal, is pushed to the far right of the image via RF cycling of adjacent k-space lines (left panel). This artefact is typically removed by blanking this line of the image (centre panel), or can be avoided via phase cycling repeat k-space lines when acquiring an even number of signal averages (right panel). In all cases, phase encoding was in the horizontal direction.

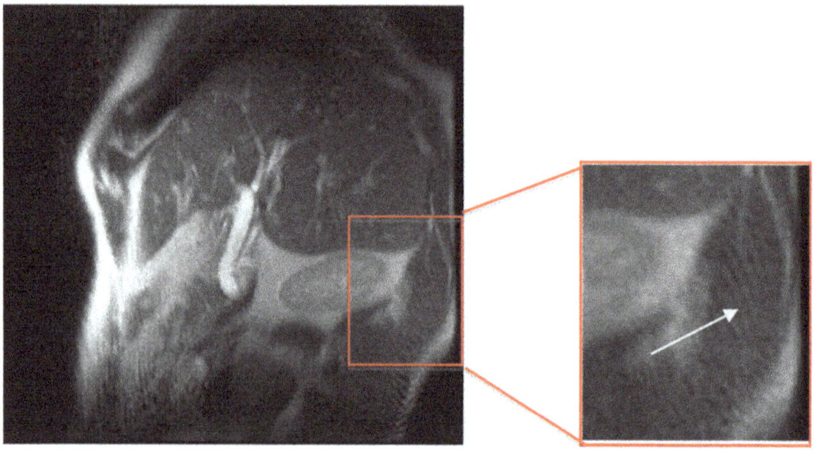

Figure 8.7 Fine-line artefact from unwanted signal that is both phase-encoded and frequency-encoded

associated with high spatial resolution and thinner slices will act to cancel this spoiling and increase the intensity of fine-line artefacts. On some systems, it is possible to increase the duration of the spoiling gradients at the user level at the expense of a longer minimum TE and increased inter-echo spacing that in turn lead to increased T_2 blurring (see Section 8.4.7). Since the amount of fresh magnetisation created by refocusing pulses reduces as the flip angle is increased towards 180°, this can be used to suppress fine-line artefacts at the expense of increased SAR.

8.2.4 Quadrature ghosting

Traditionally, the MR signal is demodulated with a reference signal before digitisation. Quadrature signal detection is the process where the received RF signal is split into two paths before demodulation, where for one of the paths, the reference signal is shifted by 90°. After demodulation, each path is processed separately to generate the complex *k*-space data. However, problems arise with this approach if there is an imbalance between the two paths, e.g. in signal gain, or if the phase shift is not exactly 90° resulting in a so-called quadrature ghost. Unlike other forms of ghosting which propagate along the phase-encoding direction, quadrature imbalance produces a ghost of the original signal reflected in both the phase- and frequency-encoding directions, i.e. rotated by 180°. With modern digital receiver chains, this artefact has been eliminated since quadrature detection is performed digitally. If this artefact does occur on older systems, it points to a fault in quadrature balance and should be reported to the service provider for correction by a service engineer. A full description of how to evaluate ghosting in MRI images is given in Chapter 3.

8.2.5 Receiver coil non-uniformity

Failure of RF coil elements in a phased-array coil may result in RF non-uniformity artefacts, apparent as a localised area of signal loss in the image associated with the position of an RF coil element, as demonstrated in Figure 8.8. Intensity uniformity correction algorithms, that are typically employed when using phased-array coils, will be able to compensate for this to a certain degree and this will become less noticeable with ever increasing numbers of RF coil elements although will impact on local SNR as highlighted in Figure 8.9. Regular coil QA plays a key role in identifying problem coil elements early so they can be addressed. Similar non-uniformities may be observed as a result of RF absorption by metal objects.

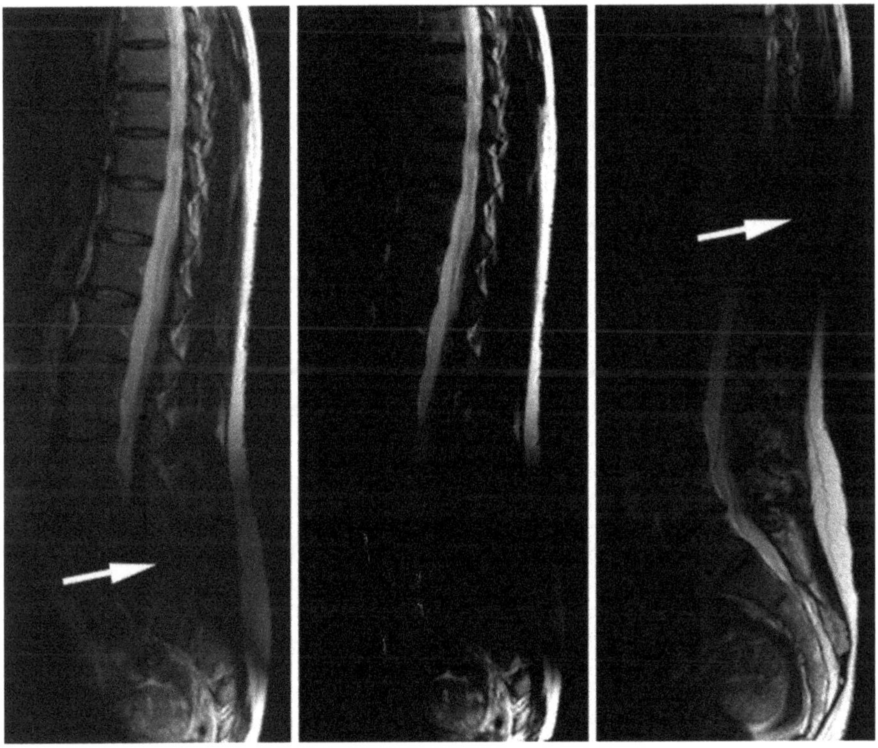

Figure 8.8 Sagittal spine images with a localised signal loss (left panel) compatible with a failure of one of the RF coil elements from a four-channel spine coil. This is more apparent upon rewindowing (centre panel). These artefacts are sometimes confused with signal loss due to metal, but repeating the acquisition with the patient turned around on the spine showed that the position of the artefact stayed with the faulty coil element (right panel)

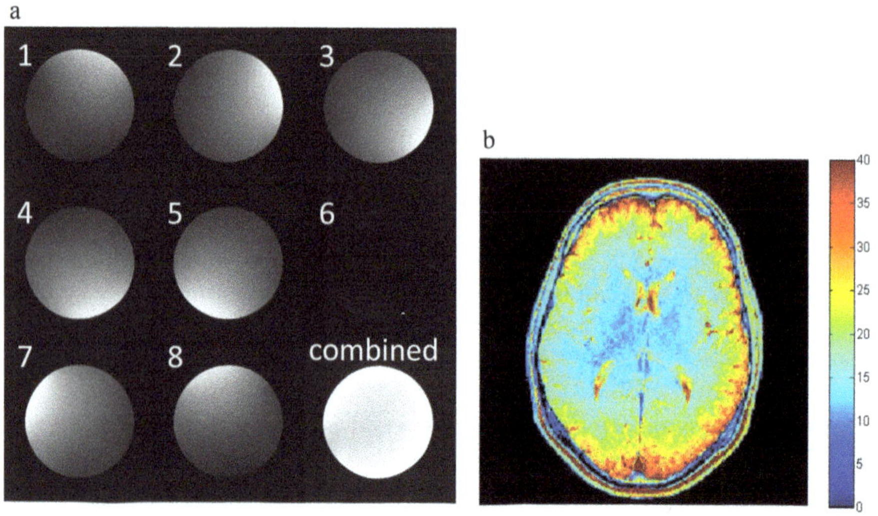

Figure 8.9 (a) Individual images from an 8-channel head coil reveal element 6 has failed. However, unlike Figure 8.8, the coil geometry and increased number of elements in this case mean the non-uniformity on the combined image is much less noticeable. (b) In vivo SNR map from this coil highlights the reduced SNR in the position of coil element 6. Images courtesy of Dr EM Tunnicliffe

As well as localised signal dropout from individual RF coil elements, faults on particular RF channels can lead to overly excessive weighting of the signal detected by the coil elements resulting in localised signal flaring, as shown in Figure 8.10.

8.2.6 Analogue-to-digital converter (ADC) overflow/signal clipping/'halo' artefact

An ADC overflow, or 'halo', artefact may result from the receiver gain being set too high so that the input voltage exceeds the maximum range of the ADC, causing the signal to be clipped. This results in signal intensity loss within the subject and apparent signal in the surrounding background, demonstrated in Figure 8.11.

Correct pre-scanning should eliminate this artefact by applying an appropriate receiver gain, and ADC overflows are rarely observed in practice. If necessary, this can be avoided by manually reducing the receiver gain.

8.2.7 B_0 inhomogeneity and gradient non-linearities – image distortions

The finite size of the magnet and the gradient coils leads to B_0 inhomogeneity and gradient non-linearity away from the magnet isocentre. These cause a breakdown in the assumed linear relationship between signal frequency and spatial location,

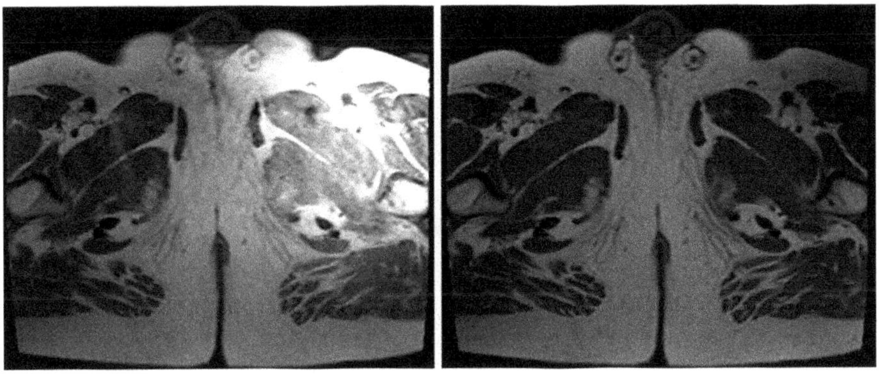

Figure 8.10 Localised signal flaring caused by a fault on a few individual RF channels (left panel). Reconstructing the same image from each channel separately identified four that had significantly higher intensities than the other 28 that contributed to the original image. Performing a retrospective reconstruction on the scanner omitting these four RF channels provided a more acceptable workaround (right panel)

leading to geometrical distortions in the image. Typically, such distortions are most noticeable at the edges of images with large FOV, particularly in the z-direction. One of the disadvantages of shorter, wider bore MR systems is the compromise on B_0 inhomogeneity and gradient linearity, although current clinical MRI systems usually provide 2D and 3D distortion correction algorithms to help compensate for gradient non-linearity, as shown in Figure 8.12.

These distortions can become more apparent when multiple stations are combined together to form extended FOV images, e.g. whole body images. The use of B_0 shimming and centre frequency adjustments to optimise the image quality for each individual station results in differences at the interface between stations that augment these geometrical discontinuities. This gives rise to the so-called broken spine artefact, shown in Figure 8.13. Additionally, different image scaling for each station can result in marked changes in signal intensity between stations, particularly between the head and chest. Some systems offer post-processing options to reduce these artefacts. Where possible, these geometrical distortions can be reduced by applying the same shim settings and centre frequency to all stations (Koh *et al.*, 2012; Li *et al.*, 2007). This may also have the advantage of avoiding additional preparation time spent running the automatic shimming routine for each station, although image quality may be compromised by non-optimal shim conditions, e.g. increased ghosting.

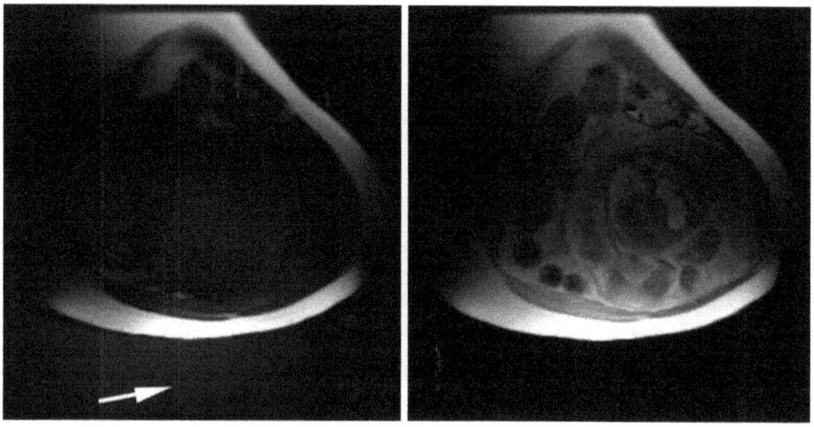

Figure 8.11 Halo/ADC overflow artefact producing washed out signal of object and signal immediately outside of object (arrow) due to receiver gain being too large (left panel). This is avoided by repeating the acquisition with a lower receiver gain (right panel)

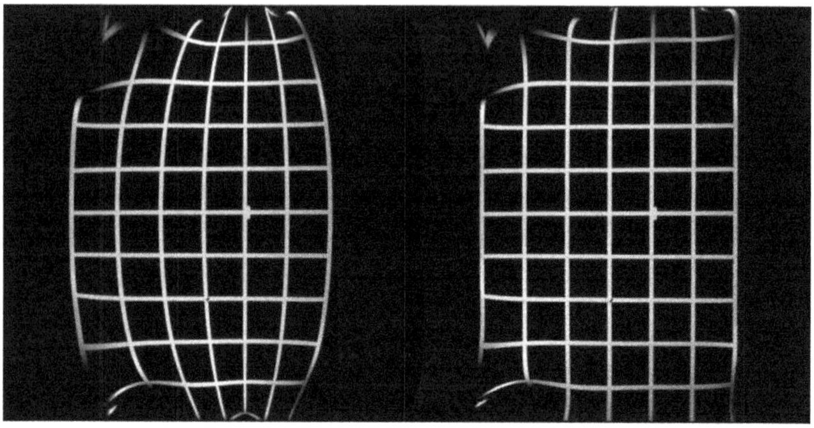

Figure 8.12 Large FOV images of a regular grid pattern phantom (spacing = 50 mm) show distortions away from isocentre as a result of B_0 inhomogeneities and gradient non-linearities that are greatest further away from magnet isocentre (left panel). Gradient non-linearities can be compensated for by the manufacturer's distortion correction algorithm (right panel)

8.2.8 Unbalanced gradients

Image-based distortions may also arise from an imbalance between the three physical MR gradients, typically as the result of miscalibration, as shown in Figure 8.14.

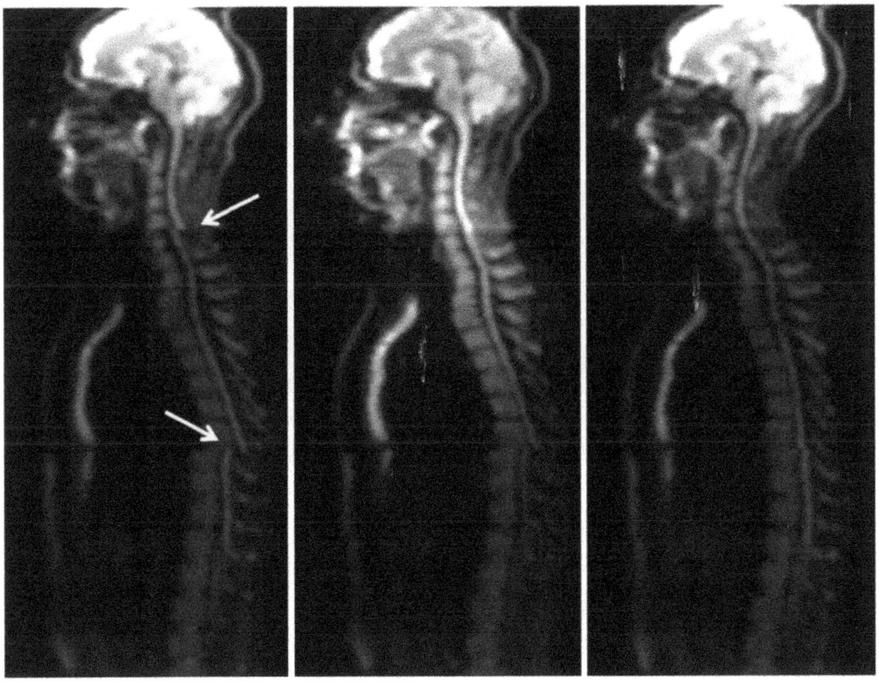

Figure 8.13 Sagittal reformat through a stack of multi-station axial diffusion images shows geometric distortions at the interface between acquisition stations leading to an apparent broken spine artefact, arising from differences in active shim settings and centre frequency (left panel). A change in image scaling, particularly between the head and chest stations, is also often apparent. Where available, the apparent broken spine artefact may be reduced by post processing options that may also attempt to normalise differences in image scaling (centre panel) or by fixing the shim settings and centre frequency for all stations (right panel)

8.2.9 Eddy currents – image distortions

The switching of the magnetic field gradients creates eddy currents in any nearby conducting material. Actively shielded gradient coils have additional wiring to help reduce the magnetic field gradients generated outside the coil, i.e. in the cryostat. Although these have long been the norm in MRI scanners, significant eddy currents are still generated inside of the gradient coil, i.e. in the RF coil. These eddy currents in turn create additional magnetic fields that persist after the switching of the primary imaging gradients has stopped and act to distort the primary gradient pulse shape. Although the impact of eddy currents can be reduced through pre-emphasis of the primary gradient pulses, residual eddy currents still persist and become increasingly significant in sequences that use large gradient slew rates and amplitudes, such as

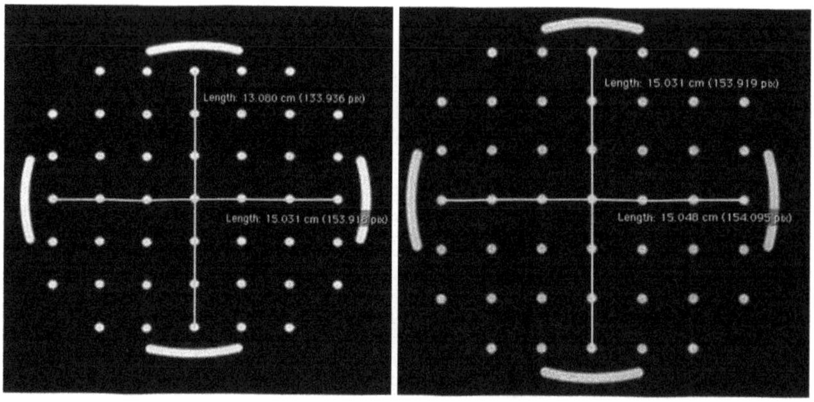

Figure 8.14 Mismatch in x and y distances of cylindrical object highlights image distortion (left panel). Following recalibration of gradients, the x and y distances are equal (right panel)

diffusion EPI, where they result in image distortions (shearing, scaling, translation), most noticeably in the phase-encoding direction due to the low receiver bandwidth.

Eddy currents arising from the diffusion gradients themselves may be mitigated through the use of twice-refocused spin-echo sequences (Reese *et al.*, 2002), although this typically increases the minimum TE. Eddy-current induced distortions depend on the amplitude and direction of the applied magnetic field gradient. Consequently, for diffusion-weighted imaging, where images are acquired using large diffusion gradients in different orientations, the resulting distortions change from one image to the next. This leads to mis-registration when the data are combined together, resulting in artefact, e.g. in the apparent diffusion anisotropy from diffusion tensor imaging, as highlighted in Figure 8.15. These artefacts can be minimised through the use of non-rigid image registration where typically, the diffusion-weighted images are registered to the non-diffusion weighted MR images.

Eddy currents can also cause a distortion of the baseline in MR spectroscopy, particularly in short TE sequences. These can be corrected by performing a repeat magnetic resonance spectroscopy (MRS) acquisition without water suppression, determining the phase distortion on the water peak and applying a correction for this to the water-suppressed MRS data (Klose, 1990), as shown in Figure 8.16. This correction can often be performed on the scanner as part of the MRS sequence.

8.2.10 Eddy currents – ghosting

As well as inducing image distortions, the time-dependent small magnetic fields created by eddy currents lead to frequency and phase shifts in the acquired data. For sequences which always acquire data with the same readout gradient polarity, this is

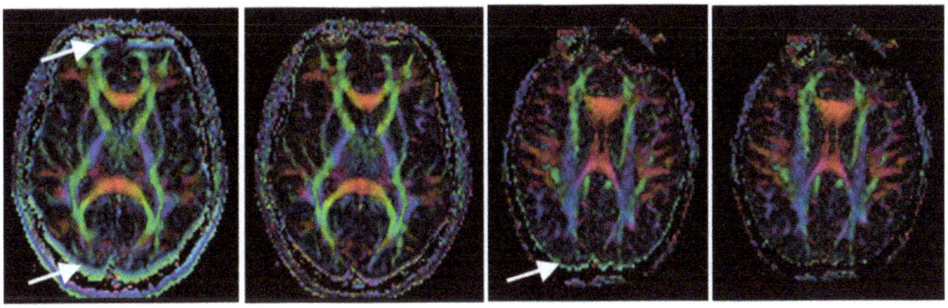

Figure 8.15 Fractional anisotropy maps can reveal high signal around the anterior and posterior edges of the object that arise from different eddy current induced distortions for each diffusion direction in a single spin-echo EPI diffusion sequence (far left panel). Phase-encoding was in the vertical direction. These distortions can be reduced during acquisition by using a twice-refocused spin-echo diffusion sequence (centre-left panel). Eddy current distortions on a different subject from a single spin-echo EPI diffusion sequence (centre-right panel) are reduced following image registration prior to DTI processing (far right panel)

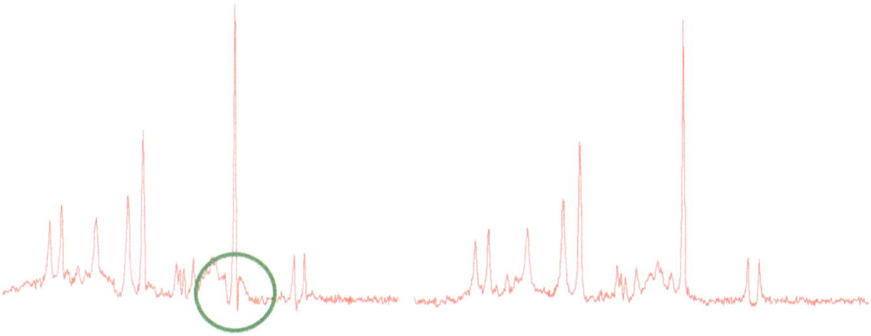

Figure 8.16 Single voxel MRS (TE=30 ms, TR=1500 ms) demonstrates baseline distortions arising from eddy current distortions (left panel), which are corrected via the unsuppressed water signal (right panel)

typically not a problem, but for EPI sequences that alternate the readout polarity between odd and even *k*-space lines, this results in a signal modulation of half the Nyquist frequency which, after Fourier transform, gives rise to a so-called N/2 ghost in the image that appears shifted from the primary image by half the FOV. EPI sequences are typically performed utilising parallel imaging which may lead to additional copies of the N/2 ghost depending on the acceleration factor, as demonstrated in Figure 8.17.

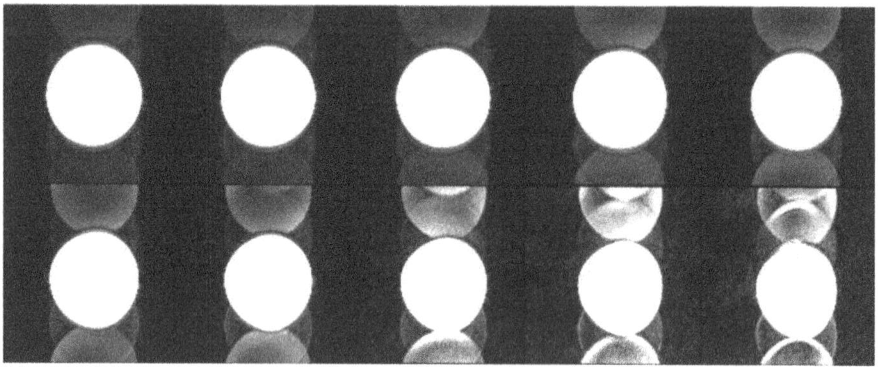

Figure 8.17 Image slices through a uniform cylinder with increasing distance away from isocentre demonstrate increasing N/2 ghosts as the magnetic field homogeneity falls off. Note some N/4 ghosting arises from the use of parallel imaging with an acceleration factor of 2. All images have identical windowing levels adjusted to highlight ghosting

In addition to eddy currents, other causes of poor B_0 homogeneity can give rise to Nyquist ghosting, including imaging away from magnetic isocentre, poor shimming due to reduced object symmetry, e.g. from a rotated head position, or a shift in patient position between shimming and imaging. Concomitant fields (Zhou *et al.*, 1998a), oblique imaging and imaging filters have also been reported to lead to Nyquist ghosting, and increased ghosting has also been observed with larger receiver bandwidths and insufficient coil loading (Delakis *et al.*, 2005).

To reduce N/2 ghosting, EPI sequences typically employ a reference scan, where data are acquired without phase-encoding to enable a correction to be made to the raw imaging data (Bruder *et al.*, 1992). Increased problems with ghosting may indicate problems with gradient timings or pre-emphasis that require recalibration and/or general shim problems. These should be highlighted to the MRI service provider to be investigated by an MR engineer. As a workaround, the FOV may be increased so that the N/2 ghost does not overlay the anatomy of interest, albeit at the expense of image quality.

Although generally less noticeable, eddy current induced ghosting can also arise in fast/turbo spin-echo sequences. On some systems, these can be corrected by turning on a phase correction option which performs as an additional pre-scan measurement to enable calculation of appropriate corrections to the RF pulse phase and imaging gradients (Ma and Zhou, 2002). Eddy currents can also lead to background phase offsets which can be suppressed on some systems by selecting a flow optimisation option that reduces the rise time of the gradients and subsequently eddy current formation, at the expense of increased minimum TE. Examples of both of these are shown in Figure 8.18.

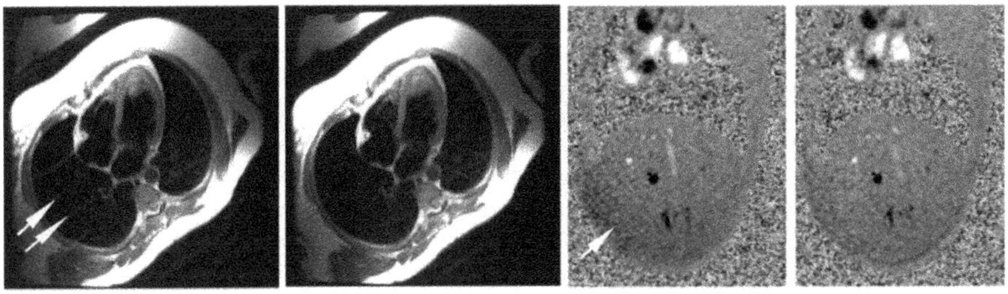

Figure 8.18 Eddy-current induced ghosting in a fast/turbo spin-echo sequence (far left panel) is suppressed when correcting the RF pulse phase and applying corrections to the imaging gradients (centre-left panel). Background phase offsets caused by eddy currents (centre-right panel) are suppressed by reducing the gradient rise time (far right panel)

8.2.11 Vibrational artefact

The large gradient pulses used in diffusion MRI cause physical vibrations. For subjects undergoing a diffusion scan, under certain conditions, these vibrations appear to give rise to low-frequency resonances in the physical structure of the scanner that are translated into propagating left–right mechanical shear-waves *in vivo* (Gallichan *et al.*, 2010), although to our knowledge, this has only been reported in the brain. These give rise to a positional shift of the signal echo in *k*-space (corresponding to a phase ramp in image space). If the shift is large enough to push the echo outside the acquired *k*-space, then a signal void will result (Meier *et al.*, 2008). This becomes increasingly more likely with the use of partial Fourier that is typically used in EPI to reduce the TE to limit distortion and T_2/T_2^* blurring. The artefact tends to be most prominent when diffusion gradients are applied in the left–right direction. When diffusion-weighted or apparent diffusion coefficient (ADC) maps are averaged over all diffusion directions, these vibrational artefacts may not be noticeable, but they become much more significant when looking at fractional anisotropy (FA) maps where they present as false areas of anisotropy as demonstrated in Figure 8.19. This further impacts on any subsequent tractography.

Reducing the use of partial Fourier and consequently increasing the proportion of *k*-space that is acquired will help to suppress this artefact, albeit at the expense of a longer TE. However, increasing the TR to allow physical vibrations to settle has also been demonstrated to reduce this artefact (Gallichan *et al.*, 2010).

8.2.12 Gradient heating

Gradient coils are non-superconducting and therefore generate heat when a current is passed through them. Furthermore, the gradients experience strong physical vibrations when pulsed within the strong static magnetic field that also lead to heating via friction. Although MRI scanners incorporate cooling systems, some heating may be transmitted to the neighbouring passive shims, the magnetic properties of which are temperature-dependent. Consequently, variations in gradient coil induced temperature may result in

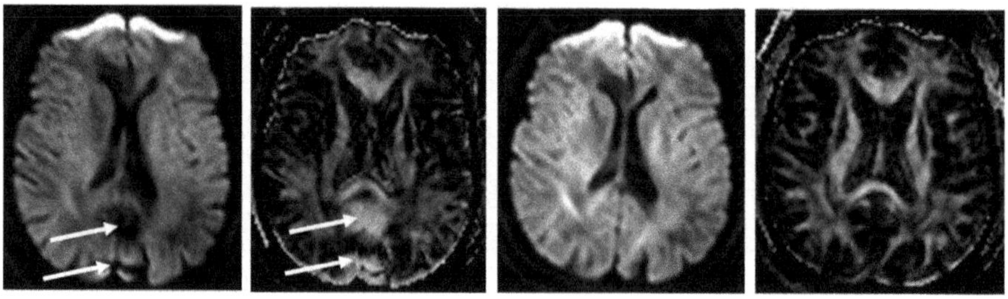

Figure 8.19 Diffusion-weighted imaging with partial Fourier can give rise to a vibrational artefact that presents as a localised signal void on images where the diffusion direction is predominantly in the left–right direction (far left panel). Consequently, this causes a noticeable artefact on the corresponding fractional anisotropy map (centre-left panel). Repeated acquisition without partial Fourier avoids this artefact and increasing the parallel imaging acceleration factor keeps a similar geometric distortion (centre-right panel). The corresponding artefact on the fractional anisotropy map is similarly avoided (far right panel)

spatial and temporal B_0 homogeneity variations during and after gradient-intensive MR sequences. These have been observed to cause a gradual reduction in frequency-selective fat saturation performance or spatial translations along the phase-encoding direction during periods of EPI imaging (Foerster *et al.*, 2005). Similar heating effects have been suggested as a likely reason for the observed drift of background offsets in phase-contrast velocity mapping on some MRI systems (Gatehouse *et al.*, 2012). To correct for frequency drifts, MRI scanners may allow frequent repeat measurements of the Larmor frequency during long gradient-intensity sequences, most notably EPI.

Immediately after heating by gradient-intensive MR sequences, similar frequency drifts will occur as the passive shims cool back to their original temperature. Significantly increased line broadening and impaired water suppression have been observed in MRS acquired immediately after an EPI sequence compared to the same measurements made before the EPI sequence (El-Sharkawy *et al.*, 2006; Lange *et al.*, 2011). It has been proposed that such line shape distortions are likely to lead to systematic quantification errors through compromised fitting. Consequently, where possible, it is advisable to avoid performing MRS acquisitions immediately after gradient-intensive imaging sequences.

8.2.13 Cusp/peripheral signal artefact

Spatial encoding of the MR signal via frequency encoding is based on the assumption of a linear relationship between the signal frequency and spatial position. Gradient coils have a limit to their physical size, and the magnetic field gradient they produce will eventually drop off. Consequently, at one end of the magnet bore, there will be a region where the magnetic field during a readout period is equal to the magnetic field within the desired imaging volume. If the MR signal from this peripheral region is excited (e.g. via coronal or sagittal slice selection) and detected (via activated peripheral receiver coil elements), then it will appear in the resulting image as a so-called cusp or peripheral signal artefact, as highlighted in Figure 8.20. When manifesting as a single point or as a ribbon of signal, this is also known as a star artefact or "annefact", respectively. This is more likely to be an issue with the shorter gradient coils found in some MRI systems. Deactivating receiver coil elements outside the imaging volume should help to avoid this problem.

8.2.14 Receiver coil decoupling failure

During the application of an RF pulse in an MR sequence, any RF coils present within the bore that are not themselves being used to generate the transmit RF field need to be decoupled. This is typically performed with the use of PIN diodes. Very rarely, problems can arise such that this decoupling fails and the RF currents that are induced in these coils subsequently generate additional RF fields which cause large local distortions in the overall B_1 field, leading to signal intensity variations within the image, as shown in Figure 8.21.

8.2.15 Other system failures

A failure of one or more components of an MR system can give rise to other artefacts, in addition to the examples listed so far. Although these issues are typi-

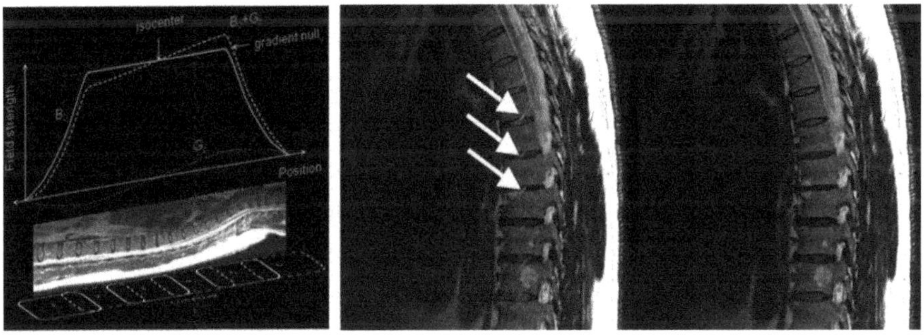

Figure 8.20 (a) Cusp/peripheral signal artefact. The additional magnetic field imposed by the z-gradient drops off at the ends of the magnet bore such that Larmor frequencies eventually match those found around isocentre. (b) If this duplicated Larmor frequency region is excited, e.g. via a coronal or sagittal slice and signal from this region is picked up by receiver coils, this will be aliased into the image (left panel). The artefact is removed (right panel) by turning off the receiver coils away from isocentre to avoid picking up signal from this region. Phase-encoding was in the head–foot direction

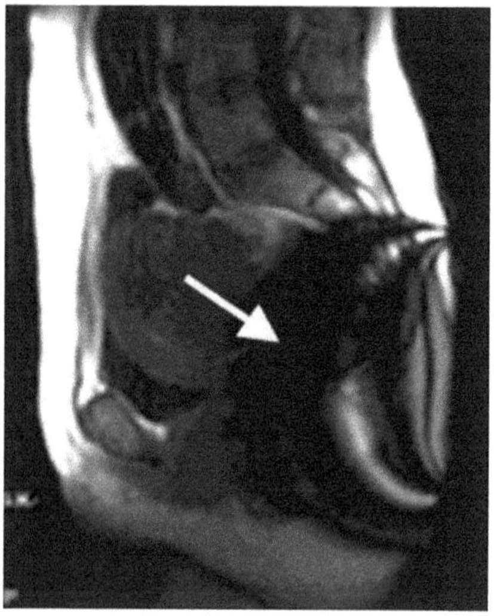

Figure 8.21 Sagittal image through pelvis showing significant localised non-uniformity due to the non-decoupling of receiver coils during the transmit RF pulse. This results in the formation of RF current within the receiver coil that leads to a secondary B_1 field from this coil producing a large variation in the actual flip angle experience in this region. Image reproduced courtesy of GE Healthcare

cally addressed by the replacement of one or more parts of the system, unfortunately, the exact cause of these artefacts is often not made apparent to the system user. Figures 8.22–8.24 demonstrate a few examples of other artefacts that can arise on clinical MRI systems.

8.3 Patient-related artefacts

8.3.1 Motion – gross patient motion

Random patient movements during an MRI acquisition generally lead to blurring in the phase-encoding direction for conventional multi-shot MRI sequences. This is due to the extended time period over which phase encoding is carried out (typically seconds to minutes) compared to frequency encoding (milliseconds). Decreasing the scan time can reduce the likelihood of the patient moving, although this may be at the expense of reduced SNR. Patient preparation is a key factor in minimising motion and this may include taking time to reassure the patient, ensuring they are comfortable and giving clear instructions before the scan and then again throughout the procedure. Sedation or anaesthesia may be required in some patient groups, and anti-spasmodic agents can be administered to temporarily minimise any peristaltic

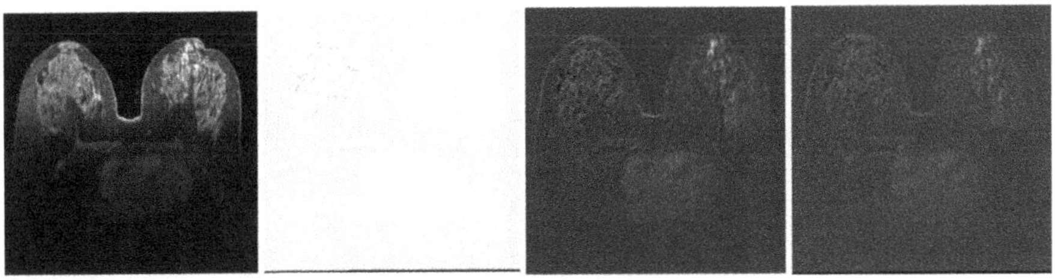

Figure 8.22 Reconstructor artefact. When scrolling through a 3D breast MR dataset, there is a sudden change from a normal-appearing image (far left panel) to the adjacent slice (centre-left panel) showing an apparent scaling issue with very high signal globally (window levels unchanged). Re-windowing this image also reveals that the image has become a subtracted image (centre-right panel). Indeed, it appears identical to the actual in-line generated subtraction image (auto windowing) (far right panel), suggesting the error occurred after generation of the subtracted image. The reconstructor on this system was replaced as part of a routine hardware upgrade and this problem was not seen again

Figure 8.23 Sagittal spine image. The marked artefactual changes were caused by an unspecified fault in the reconstruction computer. The artefacts only occurred when a coil signal normalisation filter was applied. Artefact cured by a new reconstruction computer.

movement during the scan if imaging in the abdomen or pelvis as discussed below. Additional padding, vacuum bags or compression pads may also help to immobilise the imaged area.

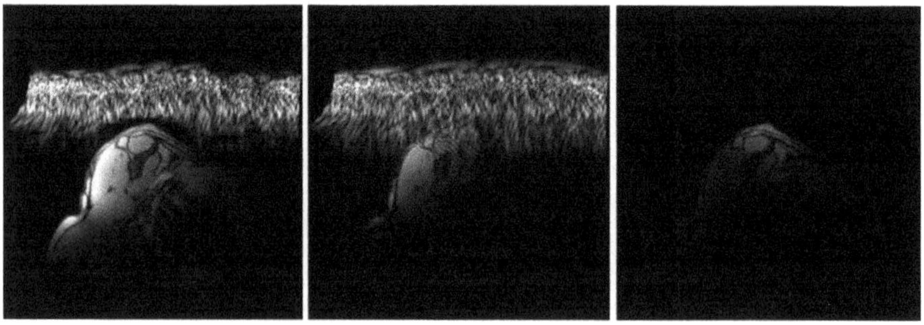

Figure 8.24 Localised band of noise in MR image from a phased-array coil (left panel). Phase-encoding was in the horizontal direction. Reconstructing this image from individual coil elements reveals that this artefact arises on some RF channels (centre panel), but others are not affected (right panel). The artefact was intermittent and the manufacturer attributed this to a poor connection within the RF coil. The coil was replaced and the artefact was no longer observed

The appearance of motion effects is intrinsically linked to the *k*-space trajectory employed. With the typical rectilinear acquisition of *k*-space, motion effects result in ghosting in the phase-encoding direction. With a radial-based trajectory, motion effects are dispersed over the different angular directions giving an effect of general unstructured blurring overall rather than distinct structured ghosts. The PROPELLER method (Pipe, 1999) is a motion correction acquisition method available now on most clinical MRI scanners (also known as BLADE or MultiVANE). This motion correction technique is based on a radial trajectory, but importantly acquires a short series of parallel lines covering the central region of *k*-space in each TR interval. Subsequent sets of *k*-space lines are then rotated by a small angle, and the process is repeated until the whole of *k*-space is covered. Because the central part of *k*-space is resampled every TR interval, a low resolution image can be generated for each set of *k*-space lines. These low resolution images can then be compared and in-plane rotations and translations between TR periods can be identified and corrections applied to the *k*-space data before it is combined with other sets of *k*-space lines. In some implementations, through-plane motions can also be identified and trigger data reacquisition. PROPELLER is effective in many clinical settings, e.g. head motion in an uncooperative patient and liver imaging where patients are unable to manage a breath hold single shot TSE sequence, as demonstrated in Figure 8.25.

If scan time permits, then a simple but often effective method to reduce motion artefacts is to increase the number of signal averages and to loop through the acquisition of *k*-space lines before looping through the repeat acquisitions (sometimes known as long-term averaging or SMART averaging), so motion effects are spread more evenly across all of *k*-space (Seitz *et al.*, 2000).

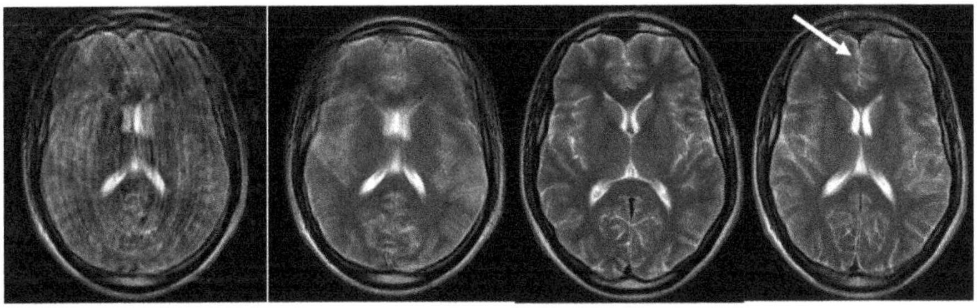

Figure 8.25 Axial brain image with subject rotating head in 'no'-type movement and a standard Cartesian-based acquisition with left–right phase-encoding (far left panel), PROPELLER type acquisition without motion correction (centre-left panel), with motion correction (centre-right panel) and with motion correction, but motion changed to a nodding 'yes'-type movement (far right panel). Note the through-plane motion still leads to some image degradation

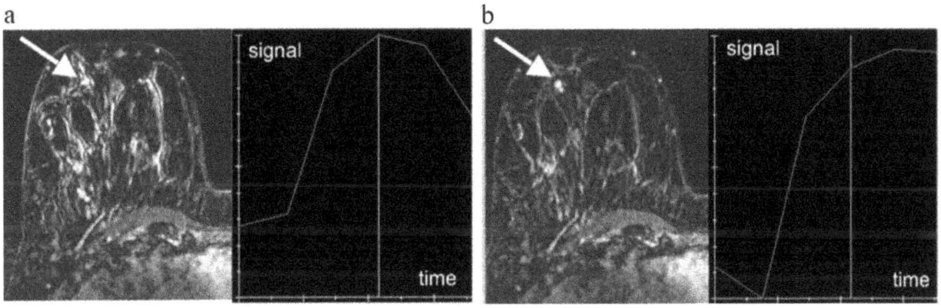

Figure 8.26 (a) Subtracted image from a dynamic contrast-enhanced acquisition (left panel) showing artefactual high signal due to motion between the two dynamic time points. The signal intensity time curve for a region of interest displays washout (right panel), more characteristic of a malignant lesion. (b) After performing a 3D non-rigid registration across the different dynamic time points, the same slice shows much less artefact (left panel) and the signal intensity time curve no longer displays wash out (right panel)

Certain MRI methods involve comparison of two or more images acquired at different time points, e.g. the subtraction of images pre- and post-administration of MRI contrast, or the analysis of enhancement kinetics over a period of time. Even though the images themselves are relatively motion-artefact free, any gross movement between acquisitions will result in errors in the overall result. Again, patient comfort and gentle immobilisation can help to avoid these problems, but if they do occur, then image registration routines available on some systems can help to correct for these problems, as shown in Figure 8.26.

In MR spectroscopy, small-scale motion during acquisition may result in varying shifts in frequency and phase across the different signal acquisitions due to magnetic field inhomogeneities (Posse *et al.*, 1993). On averaging, frequency variations will result in increased peak line-widths and changing signal phases will give rise to reduced peak areas through phase cancellation. If a single motion occurs roughly midway through the scan, then this may appear as a doubling of peaks (Kreis, 2004). To avoid these problems, frequency and phase correction of individual or small groups of signal acquisitions before averaging can be performed utilising the residual water signal (Helms and Piringer, 2001).

Larger-scale motion may lead to significant spatial misregistration, with the MRS voxel sampling different tissues away from the intended target. Acquisition and comparison of identical MR imaging through the MRS voxel location immediately before and after MRS can help to identify if any gross movement has taken place. If MRS signal acquisitions are collected before averaging, then it may be possible to identify when any large-scale motion occurred from analysis of temporal frequency and phase variations, and subsequently discard signals acquired after this point.

8.3.2 Motion – physiological (non-flow)

Involuntary motion of the internal organs can also produce motion artefacts. Such motion may be periodic, as in cardiac or respiratory motion, or random as in peristalsis of the bowel. Periodic motion leads to discrete ghosts in the phase-encoding direction. As the frequency of the motion is increased, so is the separation of the ghosts. The ghost intensity increases with the motion amplitude.

In general, physiological gating can be employed to eliminate artefacts caused by periodic motion and this typically consists of ECG or pulse oximeter gating for cardiac gating, as demonstrated in Figure 8.27, and a bellows system or navigator techniques for respiratory gating. To minimise artefacts due to peristalsis, antispasmodic drugs may be administered.

Synchronising pulse sequence repetitions to coincide with the heart or breathing cycle can introduce limitations in terms of scan time and contrast. For example, the long TR required to match the breathing cycle means T_1-weighted respiratory gated images cannot be acquired. Respiratory Ordered Phase Encoding (ROPE) provides an alternative in which phase-encoding steps are reordered such that data from the centre of *k*-space are acquired during the parts of the breathing cycle in which least motion occurs (Bailes *et al.*, 1985).

An alternative strategy is to avoid detection of the moving tissue signals altogether, for example, by using surface coils whose sensitivity drops off rapidly from the centre of the coil, or applying saturation bands, e.g. over the heart or fat areas adjacent to coil elements whose intense signal would otherwise produce stronger

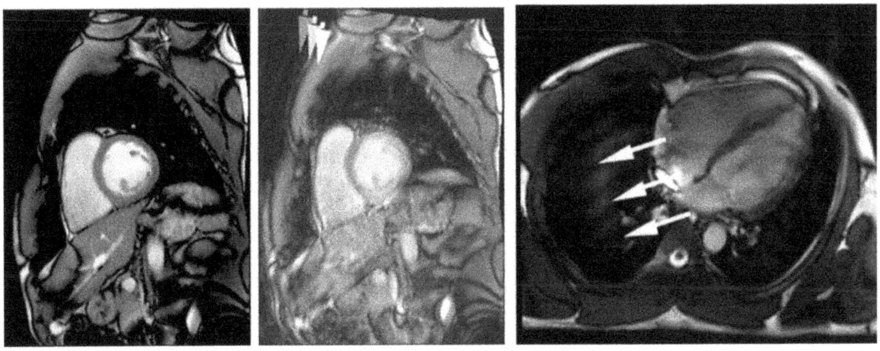

Figure 8.27 Cardiac MR examination (left panel) without and (centre, right panels) with respiratory motion artefact causing periodic ghosting

image ghosts. With modern scanners, it is increasingly becoming possible to employ fast imaging techniques such as HASTE (half Fourier acquisition single shot turbo spin echo), steady-state gradient echo, or fast/turbo spin-echo imaging together with parallel imaging to acquire images during a single breath hold.

If motion artefact cannot be avoided by the methods described above, it may be possible to minimise their appearance. In strictly periodic motion, the distance between a moving structure and its ghost is

$$d_{\text{ghost}} = TR \times N_{\text{PE}} \times NSA \times f \tag{8.1}$$

where N_{PE} is the number of phase-encoding steps, NSA the number of signal averages and f the motion frequency. Thus, it may be possible to increase TR such that the ghost lies outside the image.

Ghosting in MR cholangiography with a conventional 3D T_2-weighted fast/turbo spin-echo has been reported to be eliminated with the use of a 3D variable flip angle fast/turbo spin-echo sequence, although in this study, there were also differences in the use of parallel imaging and partial Fourier (Haystead *et al.*, 2008).

8.3.3 Motion – physiological (flow)

Flow artefacts constitute a particular subset of physiological motion artefacts. The pulsatile motion of blood within the vascular system can produce ghosting in the image. In brain imaging, artefacts due to cerebrospinal fluid (CSF) flow may also be observed.

In general, blood flow appears as ghosting of the vessel in the phase-encoding direction, as shown in Figure 8.28, with the separation of the ghosts dictated by the frequency of pulsation of flow. Flowing blood can appear artificially bright in gradient-echo sequences due to the 'in-flow' or 'paradoxical enhancement' effect in which magnetically unsaturated fresh blood enters the slice and produces a higher signal relative to static tissues that become partially saturated. Conversely, in spin-echo-based sequences, flowing blood can appear artificially dark as blood excited by one slice selective 90° pulse will have moved out of the slice plane when any refocusing pulse(s) is applied and so does not produce a spin echo. CSF artefacts can be seen, for example, as pulsatile motion artefacts in FLAIR brain images when CSF enters the slice between inversion pulse and sampling, or as bright regions in the majority of T_2-weighted brain images, most commonly in the 4th ventricle.

Saturation bands can be applied to the upstream region adjacent to the imaging volume to null the signal from blood, and hence reduce the artefact. Alternatively, flow compensation gradients can be applied following excitation to eliminate motion-induced phase differences, and hence artefacts, at the expense of an increased minimum echo time. Spin-echo sequences are less susceptible to flow artefacts due to the dark blood effect. Several techniques are available to reduce CSF flow artefacts in FLAIR images including use of 3D FLAIR, applying a non-selective 180° pre-pulse, or use of adiabatic inversion pulses.

Additional flow related artefacts may occur in time of flight MR angiography where a variation in signal saturation may result from differences in blood flow. This can occur naturally in healthy vessels due to wall friction that leads to a laminar flow profile and potential apparent narrowing of the vessel as partial saturation starts to become significant. This is more likely in large 3D imaging volumes and may be

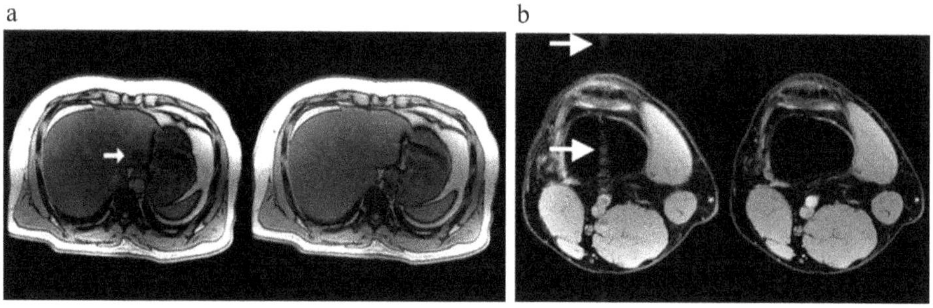

Figure 8.28 (a) Axial gradient echo image through the abdomen shows significant flow artefact (left panel) that is typically avoided by placing saturation bands immediately above and below the imaging region to suppress the blood flow (right panel). Note the reduced signal intensity in the aorta from suppressed time of flight effect. (b) Flow artefacts in a gradient echo image of the knee arising from the popliteal artery (left panel) are suppressed through the use of flow compensated gradients (right panel). In all cases, phase-encoding is in the vertical direction

avoided by utilising ramped RF pulses (Purdey *et al.*, 1992) and multiple overlapping thin slab acquisitions (MOTSA) (Parker *et al.*, 1991), although the latter of these can generate its own MRI artefact (see Section 8.4.8). Similarly, turbulent flow immediately after a stenosis may result in an apparent lengthening of the stenosis on time of flight MRA.

8.3.4 Magnetic susceptibility – normal anatomy

Changes in magnetic susceptibility create local B_0 inhomogeneities that increase the dephasing of transverse magnetisation and may result in localised signal loss, geometric distortions and variable fat suppression. In normal anatomy, significant susceptibility differences may be found at air–tissue and bone–soft tissue boundaries. It is worth noting that, in certain circumstances, anatomy-based susceptibility changes can be exploited, e.g. susceptibility-weighted imaging (Haacke *et al.*, 2004) and blood oxygenation level dependent (BOLD) MRI, the basis of fMRI.

Spin echo sequences are inherently less sensitive to susceptibility-based artefacts due to the refocusing RF pulses which help to cancel the dephasing effect of field inhomogeneity. However, signal losses will still occur with spin echo sequences since magnetic field inhomogeneities effectively sensitise the acquisition to a small amount of diffusion weighting. This can be minimised by reducing the echo time for spin echo sequences and the inter-echo time for fast/turbo spin echo sequences.

For gradient echo sequences, reducing TE and increasing matrix size can reduce the effects of susceptibility artefacts. For EPI sequences, including spin echo EPI, reducing the echo train length, e.g. through the use of parallel imaging and partial Fourier acquisition, will help to reduce the extent of susceptibility-based artefacts.

Localised magnetic field inhomogeneities may cause problems for automatic shimming algorithms that in turn may lead to image quality problems away from the area of susceptibility mismatch. Consequently, it may be advantageous to perform a localised shimming procedure with a volume defined to exclude areas containing large changes in magnetic susceptibility.

Susceptibility artefacts, as with chemical shift artefacts, are less pronounced when using high receiver bandwidths. Additionally, setting the frequency- and phase-encoding directions such that the largest bandwidth direction travels across the susceptibility mismatch is advisable to help minimise the resulting artefact. Similarly when using frequency-selective RF pulses, e.g. for saturation or inversion, the pulse bandwidth may need to be increased to cover the spread of frequencies arising from the susceptibility mismatch otherwise localised signal hyperintensities may arise, as demonstrated in Figure 8.29.

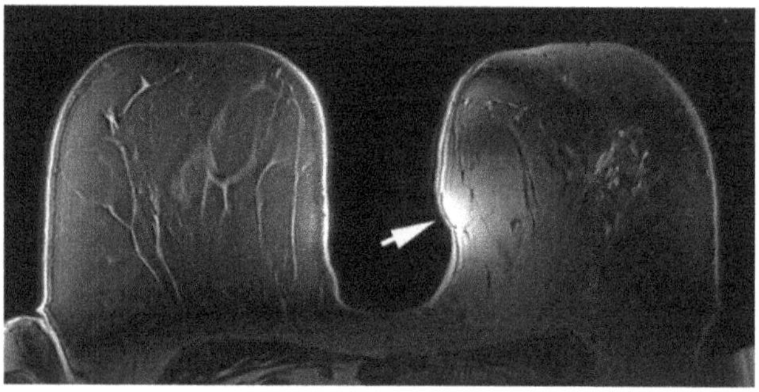

Figure 8.29 Variable fat suppression in the left breast near to air–tissue interface. Phase-encoding is in the left–right direction

Changes in magnetic susceptibility can be reduced around the area of interest by padding out difficult susceptibility regions with bags containing substances with similar susceptibility to tissue. Perfluorocarbons are often used for this purpose since they have the additional benefit of being invisible on MRI since they contain no hydrogen.

Reversing the gradient polarity will flip the direction of geometrical image distortions arising from magnetic field inhomogeneities. This can be used to visually assess the size of susceptibility-induced distortions where such issues *in vivo* are of concern, e.g. surgical or radiotherapy planning, or post-processed to construct distortion-free images (Chang and Fitzpatrick, 1992).

In localised MR spectroscopy, unwanted signal from outside the prescribed voxel is suppressed via spoiling gradients. However, partial cancellation of this suppression can arise in regions where strong local magnetic field gradients occur, e.g. from mismatches in magnetic susceptibility, leading to ghost signal in the MR spectrum. To avoid this in the brain, it is recommended that the slice-selection gradients are ordered such that the last slice is axial to avoid intersecting the sinus or basal skull (Ernst and Chang, 1996).

Magnetic susceptibility can also impact on T_1 values. The weakly paramagnetic nature of oxygen can cause a reduction in T_1 values of CSF leading to elevated signal intensities on FLAIR imaging. When using 100% oxygen as part of general anaesthesia, such effects have been observed in basal cisterns and along the cerebral sulci mimicking abnormalities such as haemorrhage or meningitis (Deliganis *et al.*, 2001; Frigon *et al.*, 2002). Although such artefacts are generally well understood by neuroradiologists, guidelines recommend that any use of 100% oxygen during anaesthesia is highlighted to the reporting radiologist (Association of Anaesthetists of Great Britain and Ireland, 2010).

8.3.5 Magnetic susceptibility – foreign bodies

Metals in general have a very high magnetic susceptibility compared to human tissue. Consequently, much larger magnetic susceptibility changes leading to more severe artefacts can occur around metal implanted within the patient, e.g. as part of medical devices, or externally as part of items of clothing, hair accessories, etc. These artefacts may extend a significant distance from the metal foreign body. Common examples include hip prostheses and dental metalwork. Metallic implants or needles may also cause artefacts as a result of changes in RF detection homogeneity due to their electrical conductivity.

Metal artefacts are manifested as areas of total signal loss, often with bright margins. Geometric distortion may be present and tissue signals may be mapped into adjacent regions or slices. The artefact appearance depends on the shape, orientation, density and composition of the foreign body and is more severe for ferromagnetic material compared to other metals, as highlighted in Figures 8.30 and 8.31. Careful screening of patients should of course be in place to ensure that those with ferromagnetic implants or other potentially hazardous implanted devices are not scanned.

The large susceptibility mismatch that arises between a bolus of gadolinium contrast agent in blood and the surrounding tissue can similarly produce signal losses in adjacent structures (Lee *et al.*, 2000), although this effect also forms the basis of dynamic susceptibility contrast MRI to provide information about tissue perfusion.

To improve time efficiency, B_0 shimming is often not performed before image acquisition if shimming over the same region and at the same bed position has already taken place. Consequently, if a patient is reimaged following removal of a metal object observed to be causing an image artefact, e.g. belt, metal fastenings on bra straps, it is important to force the MRI system to repeat the shimming process to obtain a homogeneous field, as demonstrated in Figure 8.32.

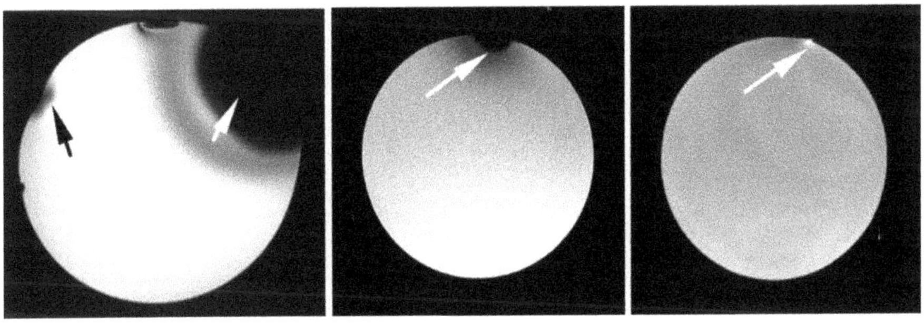

Figure 8.30 Comparison of susceptibility artefacts from a non-ferromagnetic needle (black arrow) and ferromagnetic paperclip (white arrow) at 3 T (left panel). Susceptibility artefact due to syringe needle at 1.5 T using a gradient-echo sequence (centre panel) is significantly reduced when reacquiring image with a spin-echo sequence (right panel)

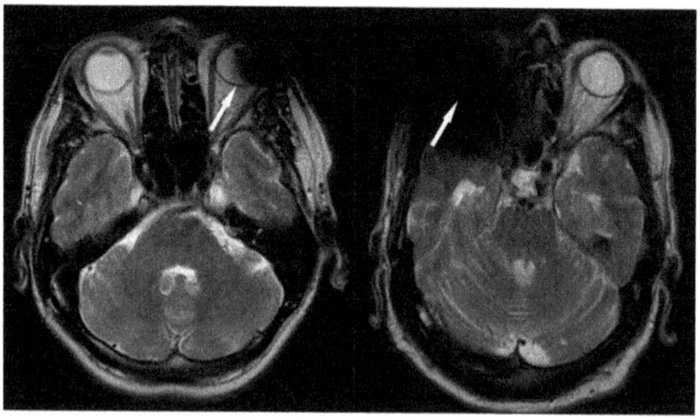

Figure 8.31 Examples of susceptibility artefact on turbo spin-echo sequences from metal particles in the eye

While it is not possible to completely eliminate artefacts from metal foreign bodies, they can be minimised through use of spin-echo sequences, high receiver bandwidths and shorter TEs. It may also be possible to swap the phase- and frequency-encoding directions to reorient the artefact away from structures of interest. For EPI sequences which are particularly sensitive to susceptibility mismatches, it is often desirable in axial imaging to have the phase-encoding anterior–posterior so that the resulting artefact is symmetrical between the left and right half of the image, as demonstrated in Figure 8.33.

Additional techniques such as View Angle Tilting (VAT), Slice Encoding for Metal Artefact Correction (SEMAC) and Multi-Acquisition Variable-Resonance Image Combination (MAVRIC) are also available on some clinical systems to further reduce metal artefacts (Chen *et al.*, 2011; Ai *et al.*, 2012; Koch *et al.*, 2009). Figure 8.34

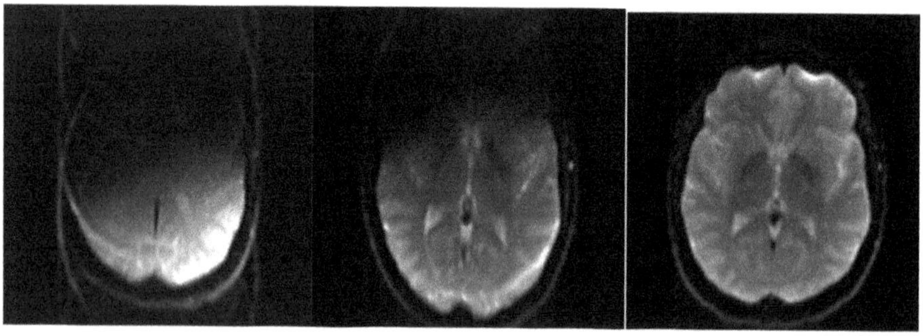

Figure 8.32 Ghosting in EPI and signal dropout caused by poor B_0 homogeneity arising from metal in a patient's bra strap (left panel). When the bra strap was moved down the arm and image acquisition was repeated, there was still significant signal dropout as the system was not aware that a reshim was required (centre panel). Only after forcing a reshim were the expected results obtained (right panel)

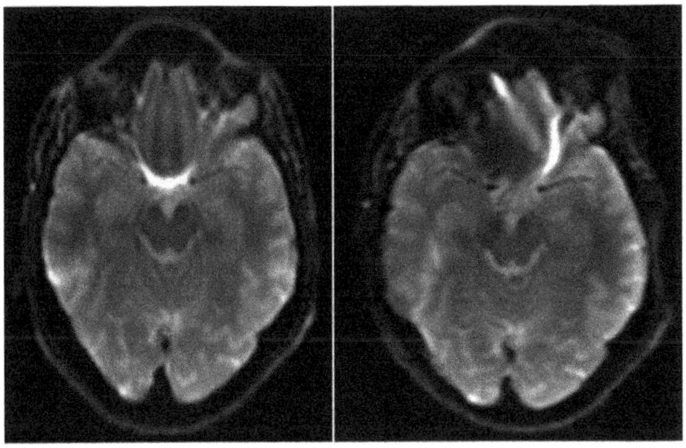

Figure 8.33 Axial EPI images are typically acquired with phase-encoding in the anterior–posterior direction which gives rise to symmetrical distortion artefacts around mismatches in magnetic susceptibility (left panel). The same image acquired with phase-encoding in left–right direction, as is common for non-EPI sequences, gives rise to a more noticeable asymmetrical artefact (right panel)

demonstrates the reduced artefact arising from the SEMAC technique in a patient with an MR conditional cochlear implant.

Even though metal objects are some way outside the field of view, they can still produce a significant image artefact through distortion of the static magnetic field,

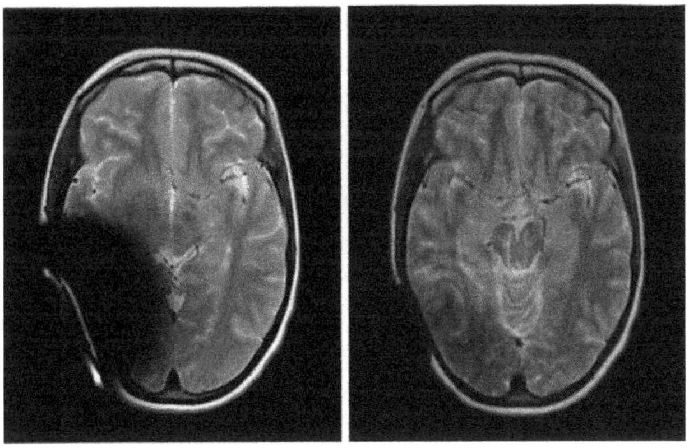

Figure 8.34 A turbo spin-echo sequence still suffers large signal dropout and image distortions from the presence of a magnet as part of an MR conditional cochlear implant (left panel). These are significantly reduced when using the SEMAC technique (right panel), although the longer acquisition time of the SEMAC sequence may lead to more motion artefacts

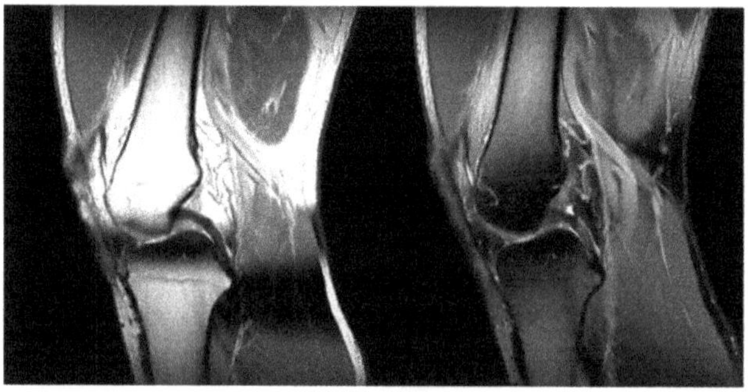

Figure 8.35 Effect of metal outside the image FOV. Poor fat suppression over the knee (left panel) is improved after the patient's belt was removed (right panel)

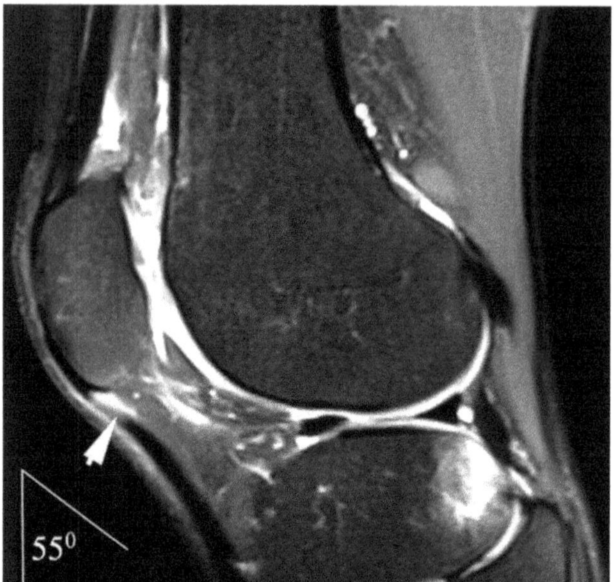

Figure 8.36 Localised increase in signal intensity of the normally dark proximal patella tendon on T_1-weighted image when at the magic angle, mimicking pathology. The magic angle is indicated in the bottom left. This localised hyperintensity was not observed on T_1-weighted axial images acquired with a longer TE. Image courtesy of Dr S Vijayanathan

as seen in Figure 8.35 where poor frequency-selective fat suppression over the knee is observed in a patient who still had their belt on during imaging.

8.3.6 Magic angle artefacts

One of the reasons that solids exhibit much shorter T_2 relaxation times than liquids is due to dipole–dipole interactions between nuclei. The strength of these dipolar interactions is proportional to $(1–3\cos^2\theta)$ where θ is the angle between the two nuclei and the direction of the static magnetic field. Importantly, this expression collapses to zero when $\theta \approx 55°$, the so-called magic angle. In liquids, dipolar interactions tend to average to zero due to rapid molecular tumbling, resulting in much longer T_2 relaxation times. The molecular motion of solids, however, is limited by the more rigid structure resulting in much shorter T_2 values.

Collagenous fibres in tendons and cartilage usually exhibit the typically short T_2 values of solids such that no signal is observed even on short TE sequences. However, since these collagen fibres are highly structured, when orientated at the magic angle with respect to the direction of the static magnetic field, these dipolar interactions reduce to zero. Consequently, the T_2 values increase such that the signal from collagen is observed on short TE sequences as focal enhancement that may resemble tissue damage, as demonstrated in Figure 8.36. However, even at the magic angle, collagen has a relatively short T_2 compared to other tissue and so slightly increasing the TE will help to suppress this so-called magic angle artefact. Musculoskeletal radiologists will typically be aware of this issue, but repositioning the subject so that structures of interest are not at the magic angle will avoid this effect.

8.4 MRI physics limitations

8.4.1 Truncation artefacts/Gibbs ringing

In MRI, the signal is digitised using a finite number of samples. If the MR signal has not completely decayed away at the edges of k-space, then Fourier truncation artefacts, also

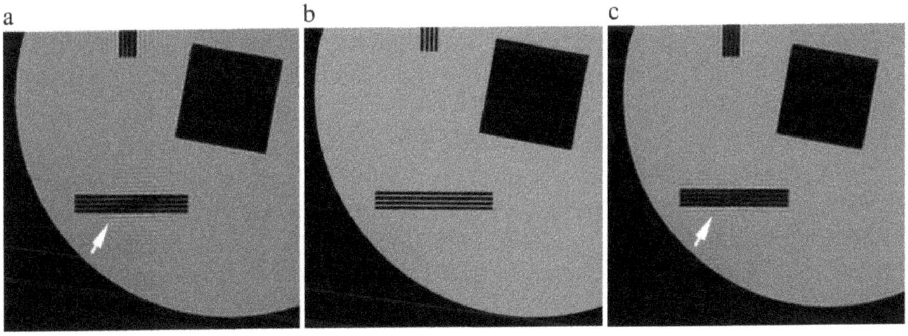

Figure 8.37 Section of the MagNET resolution phantom showing strong Fourier truncation artefact/Gibb's ringing with a 256 × 256 matrix (left panel). Increasing the matrix to 512 × 512 significantly reduces the artefact (centre panel). Reduced artefact is observed at 256 × 256 with k-space filtering (right panel). Window levels are the same for all images, chosen to highlight the artefact

known as Gibbs ringing, will appear in the reconstructed image. The presence of signal at the edges of *k*-space corresponds to high spatial frequency information in the image, associated with sudden changes in signal intensity, e.g. edge of brain, spinal cord, or interfaces between water and Perspex in a phantom. It is often more pronounced in the phase-encoding direction since zero-filling of outer *k*-space lines in the phase-encoding direction is often used to reduce acquisition times at the expense of sampling these high spatial frequencies. The most effective way to reduce the appearance of truncation artefacts is to increase the acquisition matrix size. It is also possible to apply a *k*-space filter to suppress the signal at the outer edges of *k*-space to reduce truncation artefacts. Both of these approaches are demonstrated in Figure 8.37. For some images, the edges of *k*-space will be dominated by noise and so such filters will also increase the overall image SNR, although this will always be at the expense of some loss in actual spatial resolution. Such filters are applied by default on some clinical MRI systems.

Truncation artefacts can occur whenever Fourier encoding is used. Consequently, similar issues apply to the slice-direction for 3D MRI. Again, these will be more of an issue with a lower through-plane matrix size, i.e. low numbers of slices. Such through-plane truncation artefacts are generally more noticeable when reformatting the imaging volume in orthogonal planes. Similarly, truncation can be an issue in MR spectroscopy, both in the time domain when the signal acquisition window ends before the MR signal has completely decayed away, and in the spatial domain in MR spectroscopic imaging which typically utilises much smaller matrix sizes than regular MR imaging. In MR spectroscopic imaging, the spread of signal between neighbouring voxels via Gibbs ringing is also known as Fourier bleed. A similar *k*-space filtering technique may be used in MRS, often known as apodisation.

8.4.2 Phase wrap

Conventional phase wrap artefacts can occur when the detected signal originates outside of the field of view in the phase-encoding direction. These regions are excited by the RF pulse but their signals are indistinguishable from that of tissues within the field of view resulting in the signal being 'wrapped around' into the image. The position of the wrapped signal is conventionally the opposite side of the field of view in the image, as shown in Figure 8.38, although this shifts to the middle of the image with image-based parallel imaging techniques, e.g. SENSE, ASSET (see Section 8.5.1). In clinical imaging, careful positioning and centring of the image field of view around the patient is crucial to avoid this artefact, and care should also be taken to ensure there are no other MR-signal producing objects outside the FOV in any phase-encoding directions which could then be similarly aliased into the image, e.g. fluids underneath the table top or in an intravenous line adjacent to the patient as demonstrated in Figure 8.39 (left and centre panels). For 3D sequences, phase wrap can also occur in the slice direction, as demonstrated in Figure 8.39 (right panel).

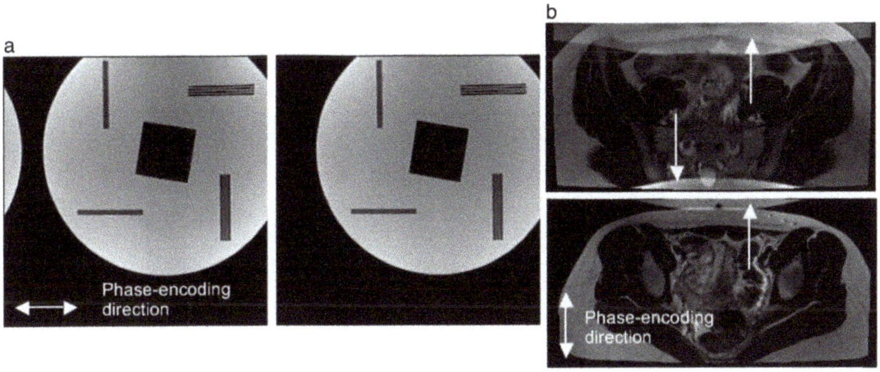

Figure 8.38 Two ways to avoid phase wrap. (a) Image of resolution test object showing wrap in the phase-encoding direction (left panel). Note the lack of wrap in the frequency-encoding direction. This is resolved with phase oversampling applied, increasing the number of PE steps from 180 to 233 (right panel). (b) Phase wrap in an axial image (top panel) is reduced by repeating the scan with a larger FOV (bottom panel)

To minimise phase-encoding steps, and therefore scan time, it is desirable to orientate the phase-encoding direction so it covers the shorter dimension of the patient's cross-section. However, consideration should always be given to the potential motion artefacts that will propagate in the phase-encoding direction for regular Cartesian acquisition, e.g. axial breast MRI are typically acquired with phase encoding left–right to keep artefact from cardiac motion away from the tissue of interest. Where it is not possible (or desirable) to contain the entire patient diameter within the phase-encoding direction, phase oversampling can be employed, adding extra phase-encoding steps to correctly encode the signals from tissues immediately outside the

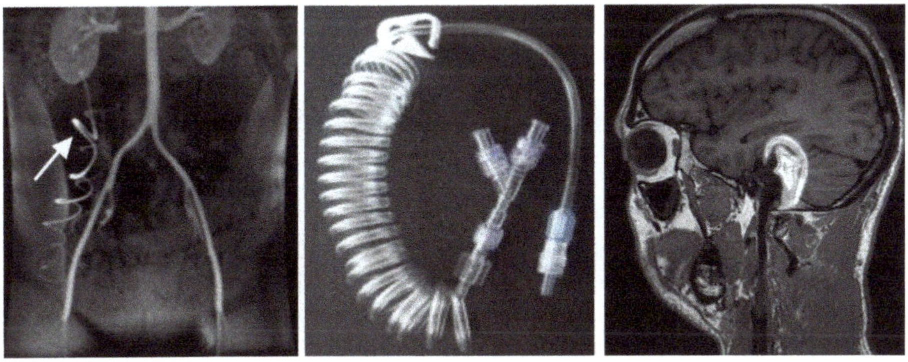

Figure 8.39 Phase wrap from signal arising outside the patient (left panel) from fluid in an intravenous line (centre panel). An ear appears in the centre of the head due to phase wrap in the slice-direction in a 3D sequence (right panel)

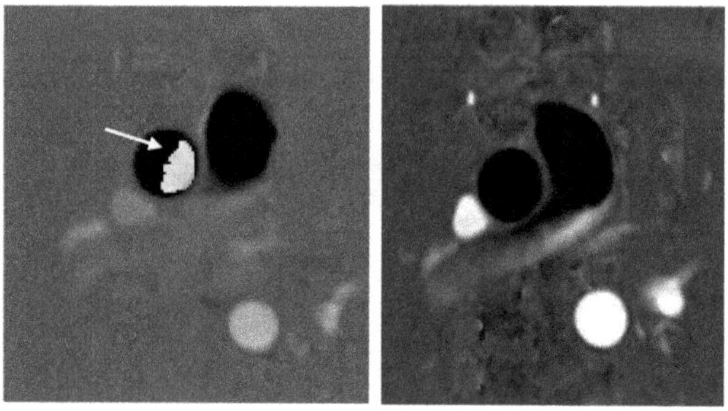

Figure 8.40 Phase wrap due to the velocity exceeding the user-defined velocity encoding parameter (VENC) in a cine phase-contrast acquisition (left panel) is resolved by increasing the VENC (right panel)

planned image field of view. Data from these regions are then discarded before image storage and display. An alternative method is to use transmit–receive coils sensitive only to a localised region, with minimal excitation and detection of signal from neighbouring regions, e.g. a transmit–receive knee coil.

In phase-contrast MRA, the user defines the velocity encoding parameter (VENC) which describes the velocity up to which the MR sequence will correctly encode for. Above this velocity, the MR signal will be aliased which will appear in the image as a sudden change in intensity from dark to light or vice versa, as demonstrated in Figure 8.40.

8.4.3 Cross-talk/saturation

Due to the finite RF pulse duration, MR scanners produce an imperfect (non-rectangular) slice profile that leads to excitation of signal immediately outside the

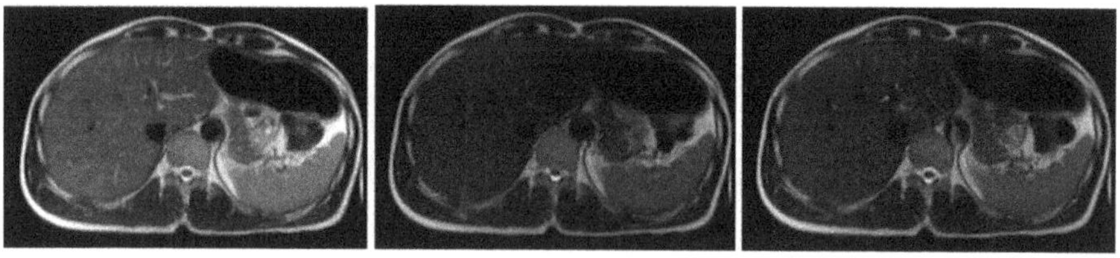

Figure 8.41 A liver image from sequence with slices acquired in interleaved order (left panel). The same sequence repeated (centre panel) with slices acquired in ascending order shows reduced signal intensity as a result of cross-talk. This is slightly compensated for by using a slice gap of 10% (0.7 mm) (right panel). Window levels are the same for all images

prescribed slice. If this outer-slice excitation falls within the region of an adjacent slice, it may give rise to partial saturation of that slice, resulting in a reduced signal intensity. This is also known as cross-talk (Figure 8.41). Saturation artefacts can also occur where oblique sets of slices at different orientations overlap, resulting in a dark band in the overlap region (e.g. in multiple axial sections through the spine). These artefacts tend to be more of a problem in spin-echo and inversion recovery sequences than in gradient-echo sequences due to the larger flip angles used. Cross-talk can be significantly reduced by using interleaved slice acquisition. Oblique sets of slices should be planned carefully to ensure they do not overlap.

8.4.4 Chemical shift

Differences in precessional frequencies due to chemical shift can lead to a spatial displacement of MR signals known as a chemical shift artefact (of the first kind). In MRI, this is most commonly observed *in vivo* at boundaries between lean tissue and fat, since the main resonant frequency for H-nuclei in fat is 3.4 ppm lower than for water, a 220 Hz difference at 1.5 T. Depending on the polarity of the imaging gradients and assuming a centre frequency matched to water, the fat is shifted in the image either towards the water, leading to an overlap of fat and water signal and a false band of signal hyperintensity, or away from the water, resulting in a false signal void, as demonstrated in Figure 8.42. The size of the chemical shift artefact is

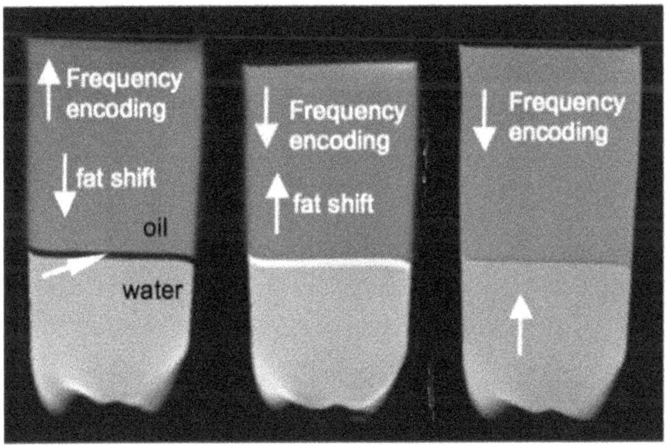

Figure 8.42 Bottle containing water and oil, imaged at 1.5 T with a receiver bandwidth of 50 Hz/pixel giving a fat–water shift of just over 4 pixels in the frequency-encoding direction. Fat has a lower frequency than water and, as such, is shifted in the opposite direction to the frequency-encoding gradient, resulting in either a signal void (left panel) or an overlap with water at the fat–water interface (centre panel). A significantly higher receiver bandwidth of 500 Hz/pixel results in a fat–water shift of less than 0.5 pixels (right panel)

proportional to the magnetic field strength and inversely proportional to the receiver bandwidth. For a conventional, non-EPI imaging sequence, each point along the phase-encoding direction is acquired at the same time offset with respect to the transverse magnetisation being in phase, either immediately before the excitation RF pulse or to the point where a spin echo occurs. Consequently, this effectively translates into an infinitely large bandwidth in the phase-encoding direction and so a zero chemical shift offset. The chemical shift artefact in the frequency direction will be dependent on the receiver bandwidth, which is user-defined. At 1.5 T, a receiver bandwidth of 220 Hz/pixel will result in a fat–water chemical shift artefact of 1 pixel. Increasing the bandwidth will reduce the chemical shift artefact at the expense of SNR. Alternating the polarity of the read gradient to flip the direction of the chemical shift is user-definable on some systems. On other systems, it may be performed indirectly by rotating the phase-encoding direction by 180°, e.g. from anterior–posterior to posterior–anterior.

For EPI sequences, the receiver bandwidth is much higher than for non-EPI sequences due to the large gradient amplitudes needed to travel quickly through *k*-space. Consequently, the chemical shift artefact in the frequency-encoding direction is very small. Conversely, because of the switched trajectory through *k*-space, each point along the phase-encoding direction is now acquired at very different times with respect to the transverse magnetisation being in phase. As a result, the bandwidth in the phase-encoding

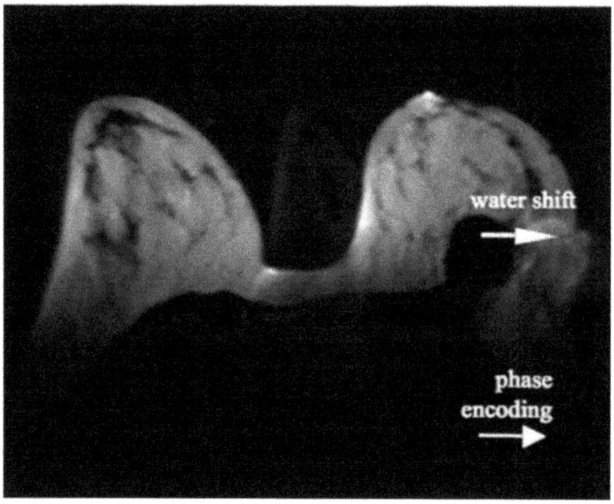

Figure 8.43 An EPI sequence shows much larger chemical shift in the phase-encoding direction. In this case, the system has set the centre frequency to fat rather than water resulting in the failure of the frequency-selective fat suppression and the water signal is shifted in the direction of the phase-encoding gradient by about 1/8th the FOV

direction is very small leading to a very large chemical shift artefact of many pixels, as demonstrated in Figure 8.43. Consequently, EPI sequences are typically performed with fat suppression to avoid the chemical shift artefact. Fat suppression can also be used in non-EPI sequences allowing small receiver bandwidths with the benefits of high SNR without the associated large chemical shift artefact.

Typically, the centre frequency is set by the MR system to match the Larmor frequency of water, which is performed by looking for the frequency with the largest signal amplitude. However, this approach is prone to errors when imaging areas with similar amounts of water and fat. A good example of this is in the breast, which may be complicated further by the third frequency of silicone in the case of implants. If the centre frequency is matched to a non-water peak, then the water signal will be shifted in the resulting image. Similarly the use of frequency-selective suppression pulses will be applied at the wrong frequency and suppression will be ineffective. Both of these effects are also shown in Figure 8.43.

There is also a chemical shift associated with slice selection, where the size of the shift depends on the bandwidth of the transmitted RF pulse. This can be relevant in MR spectroscopy since this means the actual position of the voxel varies slightly for different metabolites. Some MRI scanners can display two voxels associated with

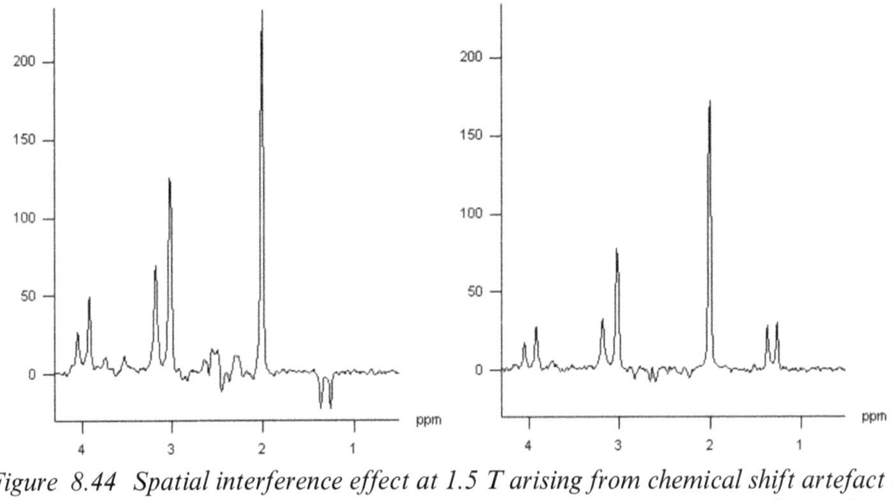

Figure 8.44 Spatial interference effect at 1.5 T arising from chemical shift artefact in single voxel MRS leads to suppression of the lactate signal at 1.3 ppm from a brain metabolite test object when TE = 144 ms, a negative in-phase TE (left panel). This suppression is minimised by acquiring at a longer TE = 288 ms, positive in-phase TE, where consequently, the magnitude of the lactate doublet is slightly increased despite additional T_2 losses (right panel)

different metabolites when planning the MRS sequence. This can be particularly helpful when positioning voxels near to areas of potential signal interference, e.g. regions of fat. Furthermore, if comparisons are planned between contralateral MRS voxels, then ideally, the frequency encoding for one voxel should be reversed to avoid the potential for misleading results due to an asymmetrical chemical shift for the two MRS voxels (Kreis, 2004).

In MRS, this chemical shift artefact also results in the so-called spatial interference effect for J-coupled metabolites such as lactate, which are then artificially suppressed although this can be avoided by utilising positive in-phase echo times with an asymmetric PRESS sequence (Yablonskiy *et al.*, 1998), as shown for lactate in Figure 8.44. A stronger effect has been demonstrated at 3 T with possible additional lactate signal losses at TE=144 ms (Lange *et al.*, 2006).

Chemical shift also leads to cancellation of signal for voxels containing similar amounts of fat and water when so-called out of phase TEs are used with gradient-echo sequences. This is sometimes known as chemical shift artefact of the second kind. The result is typically seen as a black line around different anatomical structures when surrounded by fat, as shown in Figure 8.45.

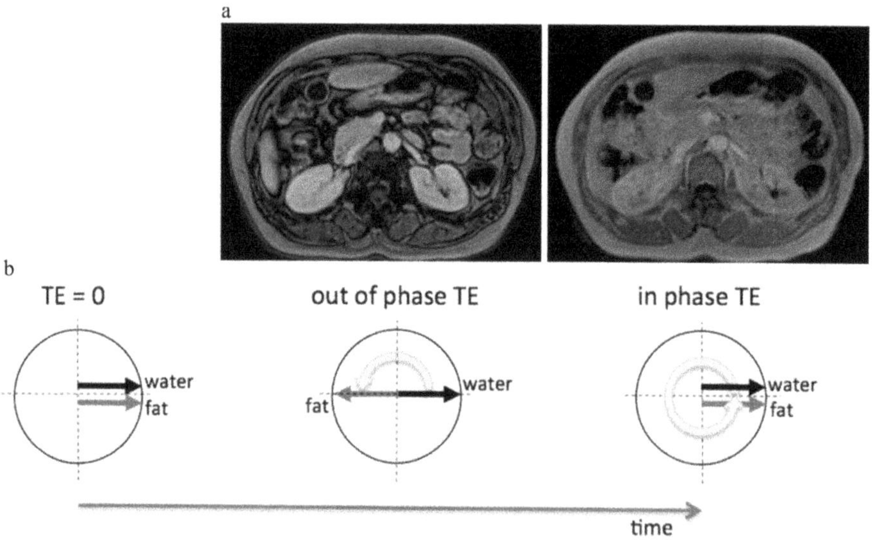

Figure 8.45 (a) Same slice through the abdomen using a gradient echo with TE producing out of phase fat and water transverse magnetisation (left panel) and in phase (right panel), the former demonstrating chemical shift of the second kind where voxels containing similar amounts of fat and water signal cancel out.
(b) Relative phases of the water and fat magnetisation against TE

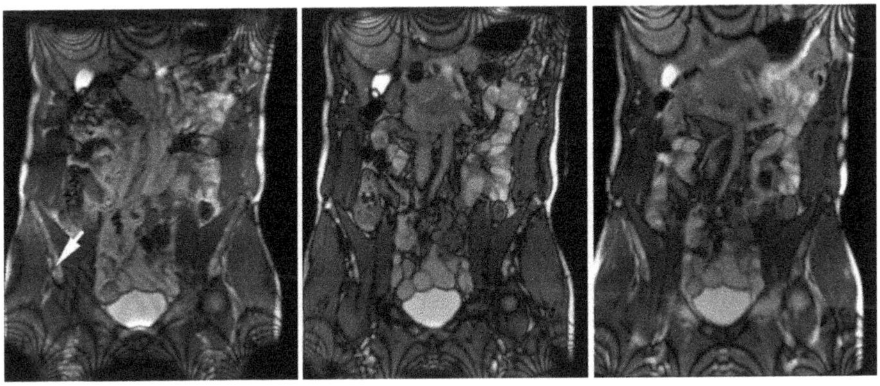

Figure 8.46 Coronal b-SSFP images through the abdomen demonstrating how banding artefacts arising from B_0 inhomogeneities change with reducing TR: TR = 10.7 ms, bandwidth (minimum) = 130 Hz/pixel (left panel); TR = 5.2 ms, bandwidth = 490 Hz/pixel (centre panel); TR = 2.6 ms, bandwidth = 1500 Hz/pixel, partial Fourier in the frequency-encoding direction (right panel). Banding artefacts decrease with shorter TR. Note addition fringe artefacts (arrow) in and increased signal dropout at extremities of image with increasing TR

8.4.5 Banding artefacts in balanced SSFP

In balanced steady-state free-precession (b-SSFP) sequences (balanced-FFE, FIESTA, True-FISP), the magnitude of the MR signal varies with the phase evolution of the transverse magnetisation between RF pulses, Φ, that arises from B_0 inhomogeneities, dropping away to zero when $\Phi = \pm n180^\circ$ where $n = 1,3,5,\ldots$ and producing a band of signal void (Carr, 1958). Consequently, reducing TR, which defines the time duration between RF pulses, will also reduce these banding artefacts, as demonstrated in Figure 8.46. Additionally, improved shimming over the region to increase the magnetic field homogeneity will also result in reduced banding.

The Constructive Interference in Steady State (CISS) technique (Casselman and Kuhweide, 1993) acquires two sets of b-SSFP images with phase cycling of the RF pulse in one acquisition resulting in mutually shifted banding artefacts that can be combined together to give band-free images, although, in practice, some residual banding may still occur, as demonstrated in Figure 8.47. However, improvements in gradient hardware mean that b-SSFP sequences can generally be performed with TR short enough such that banding artefacts may not be too problematic.

For b-SSFP sequences, as with other sequences that are sensitive to B_0 inhomogeneities, it is important to ensure the volume of imaging is centred at magnet isocentre if possible. Figure 8.47 demonstrates a case of artefact in CISS arising from B_0 inhomogeneities due

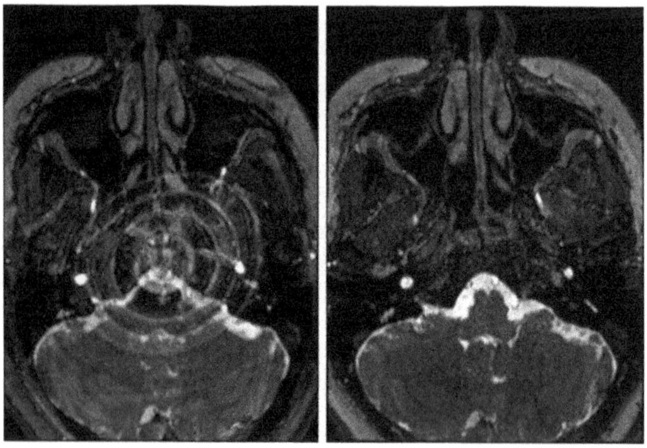

Figure 8.47 Artefacts in an axial balanced SSFP sequence through the head arising from magnetic field inhomogeneities as a result of incorrect table position (left panel) are resolved when the patient support was moved 13 cm so the imaging volume was at isocentre (right panel)

to incorrect table positioning away from isocentre. The artefact was avoided when the acquisition was repeated with the table repositioned at isocentre.

8.4.6 B_1 inhomogeneities: dielectric resonance and standing wave effects

The wavelength of an RF field is inversely proportional to the square root of the permittivity of the conducting medium, also known as the dielectric constant. Since water has a high dielectric constant, the resulting wavelength is significantly shortened to 52 cm and 26 cm at 1.5 T and 3 T, respectively. When these dimensions become comparable to the size of the imaged object, they lead to the formation of standing waves that result in variations in the applied B_1 field and hence image non-uniformities (Tofts, 1990). Oil has a much lower dielectric constant, and so is typically used in test objects at 3 T and above to avoid this effect, as highlighted in Figure 8.48.

In clinical MRI, signal non-uniformities arising from B_1 inhomogeneities are widely observed at 3 T, although more so in body imaging than neuroimaging or musculoskeletal (MSK) imaging. However, although sometimes described as dielectric shading, it appears true dielectric resonances are unlikely to be significant *in vivo* due to the heterogeneity and high conductivity of tissues that inhibit the creation of strong resonances. Instead, it appears the predominant source of these B_1 inhomogeneities comes from standing wave effects that arise from the interference of travelling waves from the different conducting elements of the quadrature body coil in the MR scanner (Collins *et al.*, 2005). These non-uniformities will depend on the pulse sequence used, since these have different dependencies on B_1 variations. Spin-echo imaging is typically more sensitive to B_1

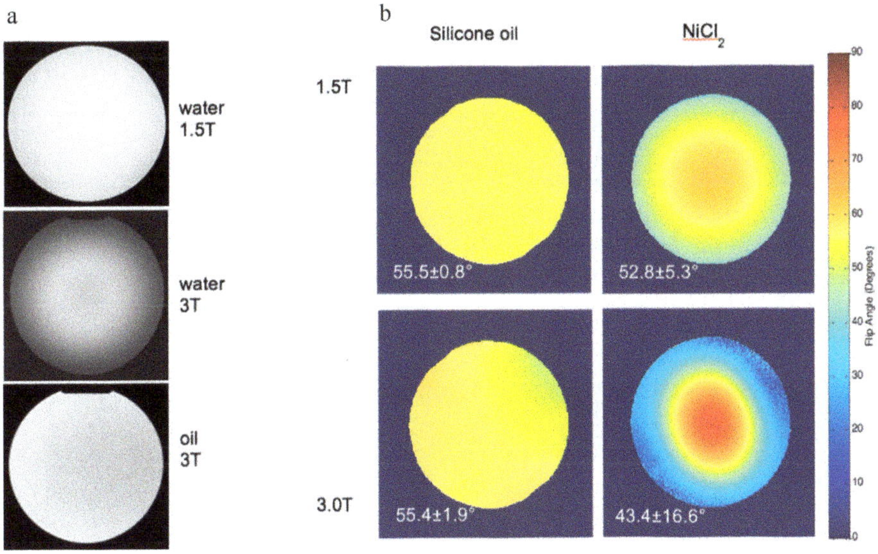

Figure 8.48 (a) A uniform 20 cm diameter water filled test object at 1.5 T (top panel) produces a marked dielectric resonance when imaged at 3 T (centre panel). No such effect is observed at 3 T with the lower dielectric constant of an oil-filled phantom (bottom panel). (b) B_1 maps produced from double angle method highlight the marked increase in the range of flip angles that result in the $NiCl_2$ water phantom when moving from 1.5 T to 3 T. Conversely, silicone oil displays much more uniform flip angle values

inhomogeneities, although turbo spin-echo sequences with lower refocusing flip angles exhibit less B_1 sensitivity.

Positioning so-called RF cushions on the patient may help to reduce this artefact, especially for thinner patients (Franklin *et al.*, 2008). However, a more recent hardware-based solution to this problem is through the use of parallel RF transmit technology, where two independent B_1 fields are applied to produce a more uniform B_1 field. This also helps to reduce local SAR hotspots that may be the SAR-limiting factor for certain sequences. Figure 8.49 shows an example of this artefact in the left lobe of the liver.

8.4.7 T_2 blurring

Fast/turbo sequences are widely used in clinical MRI. By definition, these sequences acquire multiple lines of *k*-space at different times after the same excitation pulse. Consequently, each of these *k*-space lines experience different amounts of T_2-weighting. This has a negative effect on the point spread function that leads to blurring in the image. Reducing the spread of different TEs can be achieved with a shorter

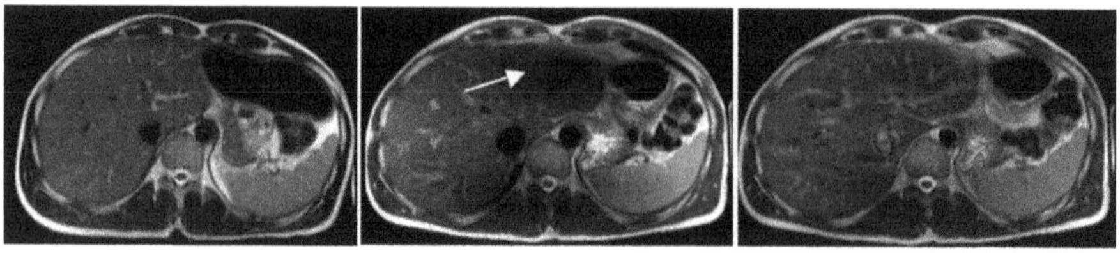

Figure 8.49 Liver images from the same subject at 1.5 T (left panel), 3 T showing reduced signal in the left lobe (centre panel) and again at 3 T utilising a 2-channel transmit RF coil to improve the RF uniformity (right panel)

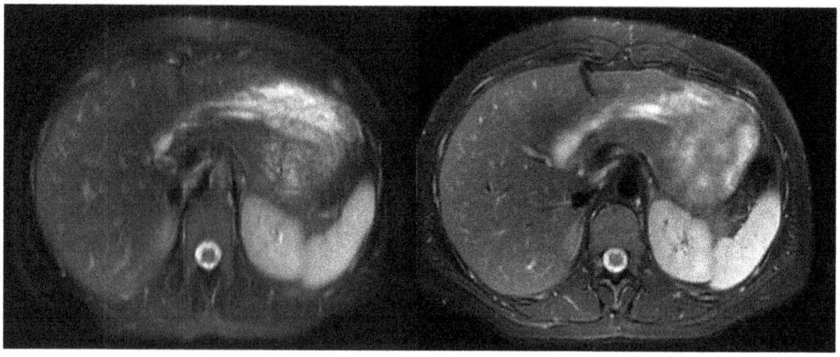

Figure 8.50 T_2-weighted image with a long echo train of 70 results in significant T_2-blurring (left panel) that is reduced by shortening the echo train length to 16 (right panel)

echo train length, as shown in Figure 8.50, at the expense of a longer scan time. Alternatively, increasing the receiver bandwidth will allow for shorter echo train spacing, at the expense of reduced SNR; *k*-space ordering schemes that fill from the centre outwards (or vice versa) will have twice the echo train spacing of linear *k*-space ordering schemes, and consequently, will display increased T_2 blurring. For 3D FSE/TSE, sequences such as Sampling Perfection with Application optimized Contrasts using different flip angle Evolution (SPACE, Siemens), Volumetric ISotropic TSE Acquisition (VISTA, Philips) and Cube (GE) use refocusing pulses with varying flip angles to counter T_2-decay and provide a more constant MR signal over a large portion of the echo train, thereby reducing T_2 blurring (Mugler *et al.*, 2000).

8.4.8 'Venetian blind' artefact

Time of flight (TOF) MR angiography relies on the saturation of stationary tissue, and refreshing of in-flowing blood provides a high MR signal. In 3D TOF-MRA, however, the blood will also experience partial saturation towards the distal end of

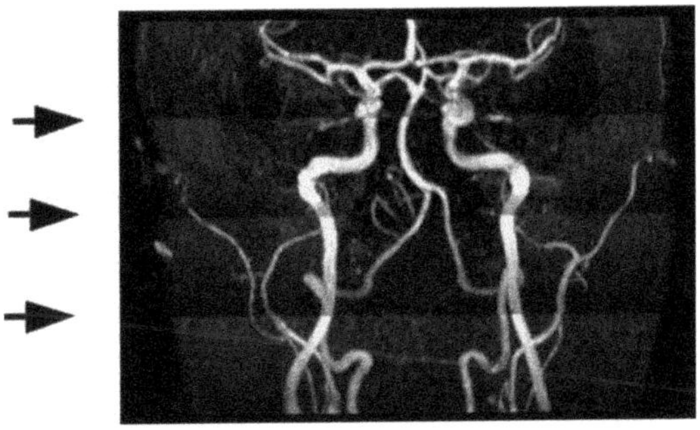

Figure 8.51 Maximum intensity projection through MRA volume acquired with MOTSA displaying a venetian blind artefact at the boundary between each sub-volume (arrows)

the 3D volume. To reduce this, a hybrid 2D–3D technique that acquires multiple overlapping thin slab acquisitions (MOTSA) (Parker *et al.*, 1991) is widely used. Ramped RF pulses (Purdey *et al.*, 1992), which counteract the increasing suppression of blood as it passes through the volume, are also typically used. This produces a gradual variation of the background signal through each slab with sudden changes at each slab interface giving rise to the so-called venetian blind artefact, as shown in Figure 8.51. This artefact is well recognised and generally not problematic, although some systems may employ post-processing techniques to suppress this.

8.4.9 Maki artefact

In contrast enhanced MR angiography, it is imperative that the acquisition of the central lines of *k*-space are timed to coincide with the first passage of the gadolinium bolus since they govern the overall image contrast. If the MRA sequence is started too early/late and the arrival of the contrast bolus does not coincide with the acquisition of the central *k*-space lines but with the edges of *k*-space, then an artificial enhancement of the edge of the vessels will occur, sometimes known as a Maki artefact (Maki *et al.*, 1996), as demonstrated in Figure 8.52. Late initiation of the MRA sequence may also result in venous enhancement.

8.4.10 Concomitant magnetic fields/Maxwell terms

Maxwell's equations state that the generation of linear magnetic field gradients always creates additional orthogonal magnetic field gradients, known as concomitant fields or Maxwell terms. These can give rise to spatial frequencies that are a non-linear function of position, resulting in positionally dependent phase-shifts. In turn, these can cause signal dropout, ghosting and a spatial offset of the image

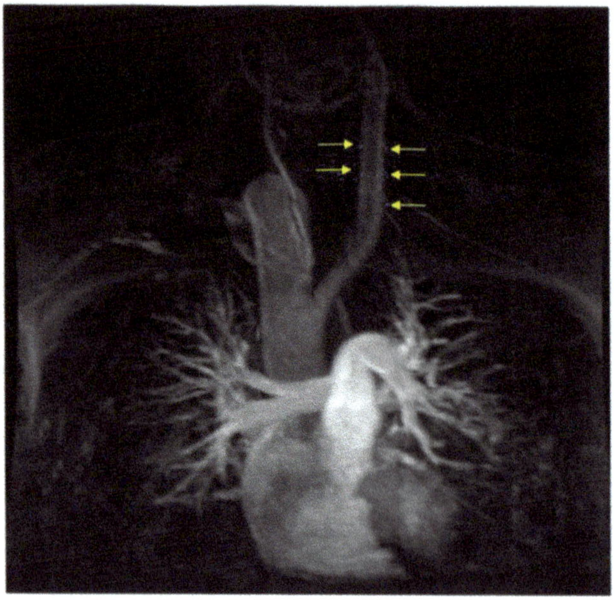

Figure 8.52 Contrast-enhanced MRA sequence was initiated too early. As a result, the contrast agent bolus was only present in the aorta and carotid when the edges of k-space were being acquired, resulting in the apparent edge enhancement, also known as a Maki artefact

which all increase as the slice position moves away from isocentre. More generally, the strength of concomitant fields is proportional to the square of the amplitude of the intended applied magnetic field gradient and inversely proportional to the static magnetic field strength. Although, at 1.5 T, the effects of concomitant magnetic fields are generally very small, artefacts resulting from concomitant magnetic fields have been observed in EPI (Zhou *et al.*, 1998a; Du *et al.*, 2002) and FSE sequences (Zhou *et al.*, 1998b). A more detailed review of concomitant magnetic fields and various methods for their correction can be found elsewhere (Zhou, 2004). These corrections are generally implemented at the sequence level. Consequently, the only course of action available to the general user to minimise any concomitant field-based artefact is to repeat the acquisition at isocentre.

8.5 Reconstruction-based artefacts

Modern MRI systems incorporate an increasing number of advanced image processing techniques that provide equal or better image quality from increasingly less acquired data. However, these techniques may introduce new artefacts, particularly when the data acquired are insufficient.

8.5.1 Parallel imaging aliasing

Conventionally, acquiring only every other line of *k*-space results in an image with half the FOV. Typically, this reduced FOV is smaller than the object resulting in image wrap, also known as aliasing, as described previously in Section 8.4.2. The central concept of parallel imaging is to utilise the different sensitivity profiles of each element in a phased-array coil to produce an unaliased full FOV image.

The so-called acceleration factor describes the reduction factor of acquired *k*-space lines and often translates into the potential reduction in acquisition time. Typically, for parallel imaging to work, it needs at least as many coil elements in the direction of acceleration as the applied acceleration factor. As the acceleration factor in parallel imaging is increased, the image reconstruction is performed with increasingly less acquired MR data. Inevitably this leads to a general reduction in image SNR, but this becomes more severe due to ill-conditioning in the reconstruction process leading to noise amplification. Typically, this occurs in the centre of the phased-array coil where the sensitivity profiles from different coil elements are least different. The spatially dependent noise enhancements are described by the so-called geometry-factor (g-factor) which is itself dependent on the phased-array coil and the acceleration factor. When scanning a 3D sequence, better g-factors are typically achieved by spreading the acceleration factor over both phase-encoding

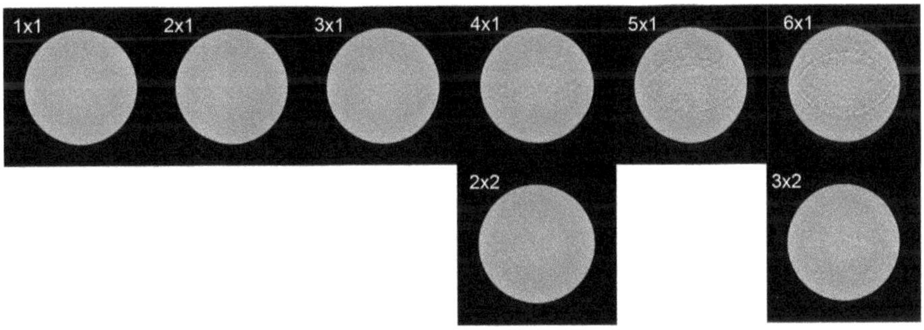

Figure 8.53 Central slice from a 3D image acquisition of a uniform test object acquired with an 18 channel head coil. The top row displays images acquired with increasing parallel imaging acceleration factor from 1 to 6 in a single phase-encoding direction (horizontal axis) demonstrating introduction and increasing artefact in the centre of image at higher acceleration factors. The bottom row shows the result when parallel imaging has been applied in both phase-encoding directions (additionally in the slice-encoding direction), with acceleration factors of 2 × 2 and 3 × 2 giving total accelerations of 4 and 6, respectively. These demonstrate reduced artefact when spreading the acceleration factor over both phase-encoding directions for this particular arrangement of coil elements

directions if the coil has multi elements in both of these directions (Weiger *et al.*, 2002; Ohliger *et al.*, 2003), as demonstrated in Figure 8.53.

Parallel imaging techniques typically incorporate post-processing algorithms to help reduce these aliasing artefacts. CAIPIRINHA (Breuer *et al.*, 2005), a recent introduction which attempts to control for such aliasing as part of a multi-slice acquisition, has been demonstrated to provide reduced aliasing, which permits parallel imaging at higher acceleration factors while retaining acceptable image quality.

Although MRI scanners may incorporate restrictions on using excessive acceleration factors, particularly when there are limited numbers of coil elements in the direction of parallel imaging, it is possible to get caught out. Figure 8.54 shows the result from a phased-array coil where parallel imaging was applied. However, there was effectively only one coil element in the phase-encoding direction, leading to regions of noise at the anterior and posterior parts of the image.

8.5.2 Parallel imaging phase wrap

Most parallel imaging techniques that operate in the image domain, such as Sensitivity Encoding (SENSE), reconstruct the MR image by directly inverting the

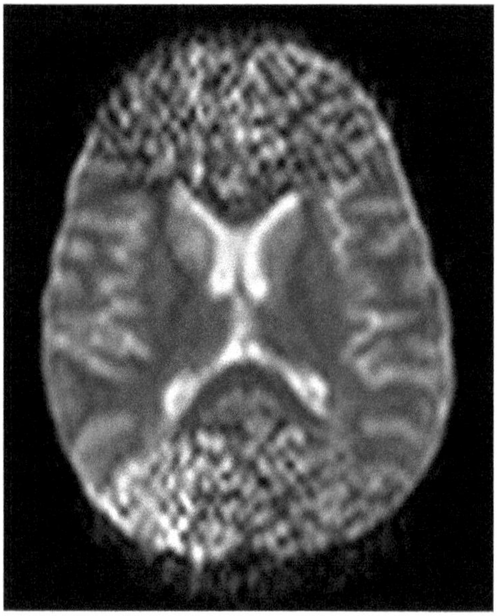

Figure 8.54 Artefact at anterior and posterior regions of head caused by use of parallel imaging with an acceleration factor of two when there was effectively only a single coil element in the AP direction

reconstruction problem, whereas *k*-space methods such as GRAPPA tend to be regenerative, working indirectly to reconstruct the missing data from data that have been acquired (Griswold *et al.*, 2004). Both approaches lead to similar results whenever accurate coil sensitivity information can be obtained, but importantly, direct methods are typically unable to deal with regular aliasing that occurs when the FOV in the phase-encoding direction(s) is smaller than the imaged object. The result of this is that part of the aliased signal appears in the centre of the image, which is generally more problematic in clinical imaging. Conversely, indirect parallel imaging methods make no assumption about the completeness of the source data and can reconstruct images with aliasing appearing only at the opposite edge of the FOV, as in MRI without parallel imaging, as demonstrated in Figure 8.55.

8.5.3 Coil sensitivity map errors

Most clinical MR images are acquired with the use of phased-array receiver coils enabling high SNR and the use of parallel imaging techniques. To get the most out of phased-array coils, spatial sensitivity maps of each element are required. These maps can be used in three ways. First, as an integral part of parallel imaging methods, second to improve upon the standard sum of squares reconstruction of the multi-coil data (Roemer *et al.*, 1990) and finally to normalise the resulting image to compensate for the inherent overall B_1 non-uniformity.

Problems arise when the position of the subject being scanned changes between the acquisition of the coil sensitivity data and the actual imaging data. This is particularly problematic for areas that have very poor information about coil sensitivity due to low MR signal, e.g. lungs. An example of the artefact that can arise when tissue moves into these regions when acquiring the imaging data is given in Figure 8.56. When the

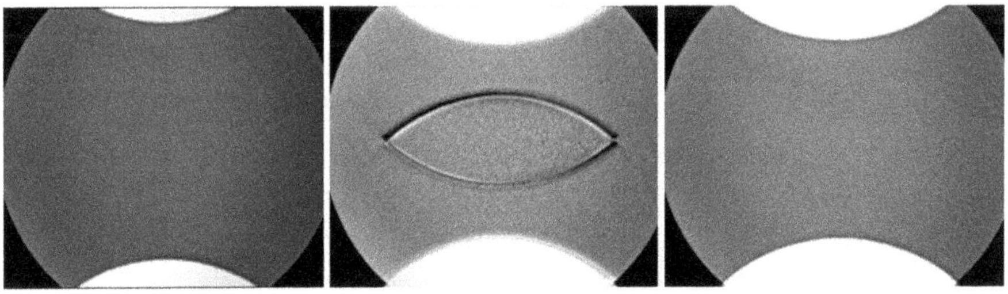

Figure 8.55 A FOV smaller than the object results in familiar wrap in the phase-encoding direction (left panel). With direct inversion based parallel imaging techniques, e.g. SENSE, the same situation gives an artefact in the centre of the image, sometimes known as "hot lips" artefact (centre panel). This can be avoided by increasing the FOV, using oversampling, or switching to a regenerative based parallel imaging technique, e.g. GRAPPA (right panel)

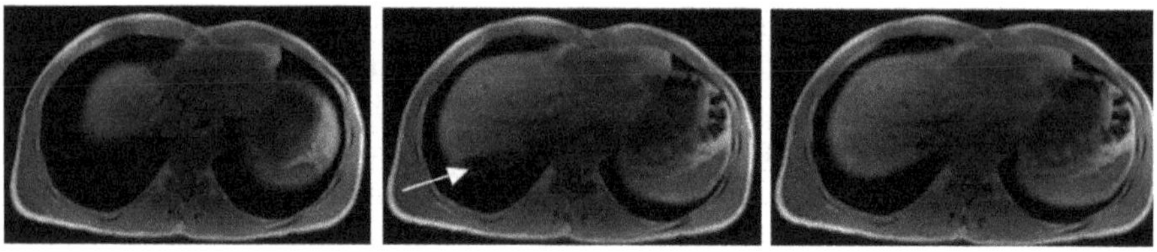

Figure 8.56 Images at the same slice position from a dynamic liver sequence reveal an inconsistent breath hold (left and centre panels). Image reconstruction is performed incorporating coil sensitivity information that is acquired from the first dynamic. Artefact on middle image (arrow) arises from poor coil sensitivity map as a result of low MR signal from this position in initial dynamic when space was occupied by lung. A retrospective reconstruction of the image on the scanner switching to a sum of squares approximation avoids this artefact (right panel)

problem arises due to integration of coil sensitivity maps into the multiple coil element reconstruction, it may be possible on some MRI systems to retrospectively reconstruct the data using the sum of squares approximation.

8.5.4 Dixon-based imaging

Dixon methods attempt to separate out fat and water signals by acquiring multiple gradient-echo images with different TEs such that the phase differences between fat and water that arise from their different precessional frequencies enable chemical shift encoding of the signal. From this, separate water-only and/or fat-only images can be calculated. Current implementations of the Dixon technique on clinical MRI systems typically incorporate iterative image processing algorithms often with region growing to allow them to be implemented as a minimum two-step sequence acquiring in-phase and out-of-phase gradient-echo images.

Occasionally, these algorithms can suffer from swaps between the calculated water and fat signals. This can happen across the whole image (global swap) such that the

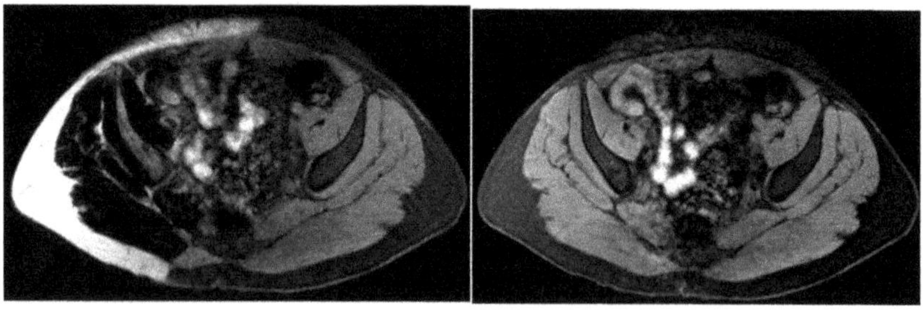

Figure 8.57 The left side of a calculated water image from a 2-point Dixon sequence (left panel) is affected by a local fat–water swap. This was avoided when repeated as a 3-point Dixon sequence (right panel)

labelled water image is actually the fat image and vice versa, or it can occur in just a portion of the image (localised swaps). In practice, even though they may not be required, it is safer to reconstruct both the water and fat images so the required images are still obtained should global fat–water swaps occur. Comparison with the acquired images may help to clarify if a localised swap is suspected. If available, moving from a two-point to three-point Dixon may help to avoid fat–water swaps, as demonstrated in Figure 8.57, at the expense of a longer acquisition time.

8.5.5 Partial Fourier

Partial Fourier describes the technique of acquiring an asymmetric area of *k*-space, with a reduced portion of data collected on one side around the centre. A more detailed description of the different partial Fourier techniques that exist can be found elsewhere (Liang *et al.*, 1992), but they essentially fall into three groups: zero filling, Margosian or homodyne reconstruction and iterative methods, all of which are currently implemented on different clinical MRI systems. Partial Fourier can often be used to reduce the impact of artefacts, e.g. to reduce echo train lengths which reduces T_2 blurring in single shot fast/turbo spin-echo sequences and allows a shorter TE in EPI sequences thereby increasing the SNR. As with most techniques that acquire less data, increasing the use of partial Fourier factor will result in a general SNR reduction, as the noise from the fully sampled side of *k*-space is copied across to the partially sampled side. However, additional image artefacts may arise,

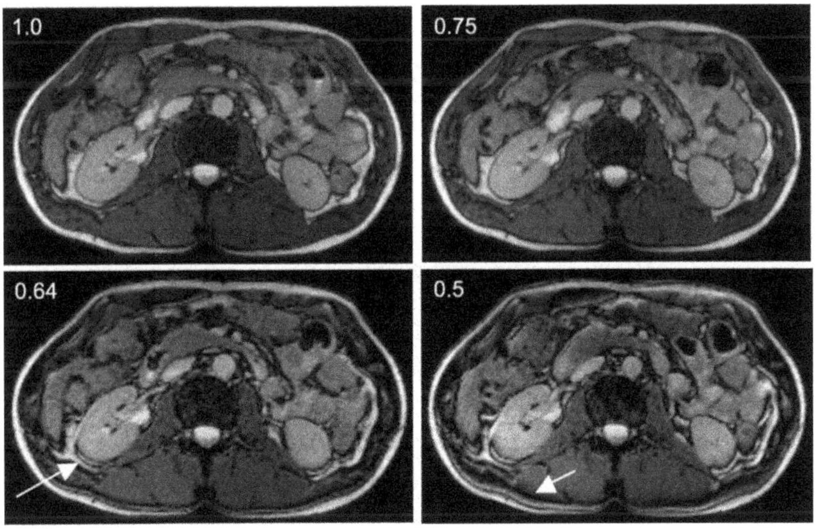

Figure 8.58 Abdominal balanced SSFP imaging with increasing amounts of partial Fourier, from 1.0 (no partial Fourier) to 0.5. The emergence of artefacts is highlighted with use of 0.64 and 0.5 partial Fourier factors

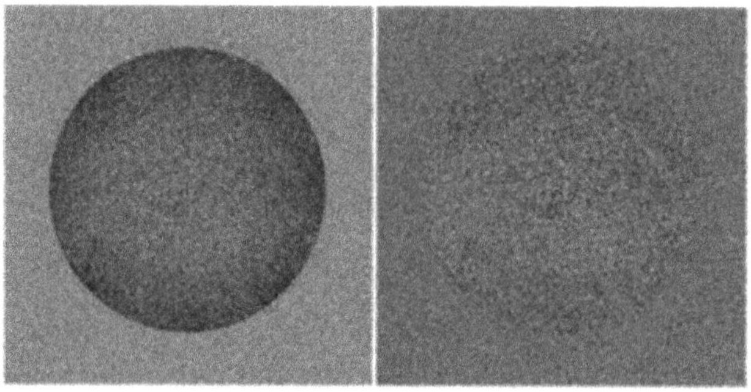

Figure 8.59 Subtracted images from a double image SNR measurement. Different image intensity scalings, if not taken into account (left panel), produce an artificial apparent residual signal. The true subtracted image is produced (right panel) when taking these intensity scalings into account. Images courtesy of Dr A McCann and Dr C McGrath

particularly where there are strong phase variations within the image. These artefacts become more prominent as the amount of acquired data is reduced, as demonstrated in Figure 8.58.

A comparison of a number of partial Fourier approaches (McGibney *et al.*, 1993) found that iterative methods generally produce a more robust result with fewer artefacts. More recently, zero padding has been demonstrated to minimise artefacts in diffusion-weighted images due to cardiac pulsatility when compared to the Margosian approach (Robson and Porter, 2005). All partial Fourier techniques will produce signal loss whenever the signal is shifted in *k*-space outside of the acquired data. This can be caused by significant phase changes arising from vibrations (see Section 8.2.11) or concomitant fields (see Section 8.4.10).

8.5.6 Intensity scaling

Acquiring repeated MR images while changing a single acquisition parameter is widely used in MRI to allow calculation of quantitative parameters such as apparent diffusion coefficient (ADC), T_1, T_2 and T_2* values. These relaxation time measurements may be utilised further to produce quantitative maps of physiological parameters such as perfusion and permeability.

When using third-party software for such processing, it is important to be aware of image intensity scaling that may be performed on some MR systems. A recent assessment of different third-party software tools found that the majority did not take these intensity scalings into account, leading to errors in the calculation of such parameter maps (Chenevert *et al.*, 2014). A simple example of this is shown in Figure 8.59 where a subtracted image calculated from a double image method as part of an SNR measurement, shows an apparent residual signal when intensity scaling is not taken into account (McCann *et al.*, 2013).

8.6 Summary

We conclude this section with a table summarising the causes and appearances of MRI artefacts plus strategies for their removal or reduction (Table 8.1).

Table 8.1 Summary of the causes and appearances of MRI artefacts plus strategies for their removal or reduction

Type	Artefact	Causes	Appearance	To eliminate/reduce/ workaround
System faults and hardware limitations	Continuous RF interference – 'zipper'	Fault in the RF cage RF from ancillary equipment Mains electricity/ faulty light bulbs Transmit RF break through	One or more lines of alternating bright/dark pixels across image in phase-encoding direction Transmit RF break through: line will be in centre of image, i.e. Larmor frequency	Ensure integrity of RF cage – check for cables left in waveguides, door open during scanning Repeat artefact-affected sequence and turn off ancillary electrical equipment in the magnet room one by one to identify source of interference Use MR compatible equipment in magnet room Check for potential RF interference from other MRI scanners nearby, e.g. magnet room doors open during RF transmit Reduce receiver bandwidth, adjust centre frequency, keep centre frequency of adjacent MRI scanners far apart
	Instantaneous RF interference – RF spike	Electrical transient (e.g. static discharge or arc) during signal collection Breakdown of RF coil	Corduroy (1 spike), 'Herringbone' pattern (>1 spike), or noise-like if many spikes	Check coil connections Check light bulbs Avoid very low humidity in scan room Repeat reconstruction after manual removal of spikes in *k*-space where possible
	DC offset/ baseline errors/ spurious signal	Constant offset in raw data – only relevant for older scanner generation Spurious/unwanted signals arising from unsuppressed signal coherences, typically free induction decay formed by imperfect RF refocusing pulses	Constant DC: bright pixel at image centre (may spread into adjacent pixels along phase-encoding direction due to temperature variations) Frequency-encoded: interference line in the frequency-encoding direction which is typically shifted from the centre to the edge of image via phase cycling between alternate *k*-space lines, or suppressed with phase cycling over even numbers of signal averages Frequency- and phase-encoded: localised area of fine-lines	Constant DC: phase cycling with even number of averages should remove Fine line artefacts reduced by increasing gradient spoiling if possible and/or increasing flip angle of refocusing pulses

Table 8.1 Continued

Type	Artefact	Causes	Appearance	To eliminate/reduce/ workaround
	Quadrature ghosting	Mismatch in gain between quadrature coil channels – only relevant for older scanner generation	Ghost of image rotated by 180°	
	Receiver coil non-uniformity	RF coil element failure Lack of uniformity correction when acquiring signal with phased-array coil RF absorption by metal objects/ implants	Loss of signal across part of image	Regular coil QC/service Turn on uniformity correction if not already Removal of metal items
	ADC overflow/ Halo artefact	Receiver gain too high causing ADC overflow	Intensity loss, apparent signal outside object	Check receiver gain/prescan Reduce receiver gain manually
	B_0 inhomogeneity and gradient non-linearity	Static magnetic field inhomogeneity Gradient magnetic field nonlinearities at edge of field of view Patient based – susceptibility	Spatial distortions, particularly away from magnetic isocentre	Ensure imaging volume is centred at magnetic isocentre Use manufacturer's gradient distortion correction schemes If sudden problem, check for shim issues, erection of metalwork near magnet room?
	Unbalanced gradients	Mis-calibration of different gradient coils	Squashed images	Engineer required to recalibrate gradients
	Eddy currents – image distortion	Eddy currents formed in conducting material (e.g. RF coil) from the switching on/off of gradients coils	Diffusion MRI: slight distortion of images that is most noticeable when comparing diffusion images from different diffusion directions, e.g. FA map MRS: distortions in baseline around peaks	Consider distortion reduction techniques, twice refocused spin-echo EPI, diffusion image registration, MRS eddy current correction
	Eddy currents – ghosting	Signal modulation at half the Nyquist frequency due to differences in odd and even k-space lines as a result of timing errors in gradient pulses, poor B_0 homogeneity, concomitant fields	Ghost shifted half the FOV with respect to object in image Additional ghosts may appear when parallel imaging is used or if using multi-shot EPI sequences	Check imaging volume is at magnet isocentre Reduce bandwidth Improve shimming Increase FOV If increase in ghosting noticed, contact MRI service engineer. Ask MRI service engineer to recalibrate eddy current compensation

Table 8.1 Continued

Type	Artefact	Causes	Appearance	To eliminate/reduce/ workaround
	Vibrational artefact	Gradient-induced physical vibrations translated into left–right mechanical waves in the brain resulting in shift of signal echo outside acquired k-space	Localised signal voids, observed when diffusion gradients are applied in left–right direction	Increase amount of k-space acquired by reducing use of partial Fourier. Alternatively increased (doubled?) TR may help
	Gradient heating	Temperature changes of passive shims from heating via pulsing of gradient coils during gradient-intensive sequence and afterwards as passive shims cool	Drift in resonance frequency, variation in performance of frequency-selective operations, e.g. fat saturation	Additional frequency adjustment measurements during periods when temperature changes most likely Avoid frequency sensitive measurements, e.g. MRS, immediately after gradient-intensive MR sequences
	Cusp/ peripheral signal artefact	Signal from region at edge of bore detected and mapped onto image	May appear as single point or a ribbon of signal	Match activated receiver coil elements to FOV
	Receiver coil decoupling failure	Detuning of receive coils fails during B_1 transmission	Localised non-uniformity around faulty receive coil	Report to MRI service provider. Take coil out of action as potential to heat up and cause patient burn
	Other system failures	Software/hardware problems	Various	Report to MRI service provider. If possible, it may be useful to retrospectively reconstruct images from individual coil elements, and/ or turning off filters to aid service provider in diagnosing problem
Patient based	Motion – gross patient motion	Uncooperative patient, inability to hold breath for duration of breath hold scan	Ghosting/blurring in phase-encoding direction	Reduce scan time (typically = trade-off with SNR) Better patient instruction, comfort Increase number of averages Sedation/compensation/ anaesthesia where appropriate Motion-insensitive/compensation techniques, e.g. radial PROPELLER type acquisition

Table 8.1 Continued

Type	Artefact	Causes	Appearance	To eliminate/reduce/ workaround
	Motion – physiological (non-flow)	Cardiac motion Respiratory motion Peristalsis	Periodic motion: Discrete ghosts Spacing ↑ as frequency ↑ Intensity ↑ as amplitude ↑ Random motion: Blurring	ECG/peripheral gating Breath-hold scanning/Respiratory Ordered Phase Encoding (ROPE) Antispasmodics to reduce peristalsis Apply saturation (SAT) bands to moving tissue Use surface coils to avoid excitation of moving tissue Select phase-encoding direction to divert artefact away from areas of interest in image Fast imaging techniques, parallel imaging Increase distance to ghost by increasing TR, number of averages, number of phase-encoding steps
	Motion – physiological (flow)	Blood flow CSF flow	Ghosting of vessel in phase-encoding direction, spacing of ghosts dictated by frequency of pulsation of flow In-flowing blood: Bright in gradient-echo, dark in spin-echo CSF artefacts bright regions in T_2-weighted brain images, commonly in 4th ventrical	Blood flow: SAT bands adjacent to imaged · volume Flow compensation gradients (increases minimum echo time) Spin-echo sequences less susceptible to flow artefacts due to dark blood CSF: 3D FLAIR Non-selective 180° RF pulses Use of other sequences/planes
	Magnetic susceptibility – anatomy	Tissue/air boundaries Tissue/bone boundaries Haemosiderin component in blood	Signal loss, particularly at air–tissue boundaries, bone–soft tissue boundaries Geometric distortion may be introduced around interface Ghosting in MRS	Use spin-echo sequences in preference to gradient-echo Reduce TE/use segmented acquisition Increase matrix size Patient-specific shim Avoid phase-encoding direction perpendicular to boundaries with susceptibility differences MRS: avoid last slice selection gradient going through area of susceptibility

Table 8.1 Continued

Type	Artefact	Causes	Appearance	To eliminate/reduce/ workaround
	Magnetic susceptibility – foreign body	Large magnetic susceptibility of metals (implants, needles, etc) causing field distortions	Areas of total signal loss, bright margins, distortion. Tissue signals may be mapped into adjacent areas or slices Severity dependent on shape, orientation, density and composition of material More severe for ferromagnetic materials	Thorough patient screening for internal metal object if scan is non-diagnostic Use spin-echo, fast/turbo spin-echo sequences Use high bandwidths Use shorter TEs Avoid frequency-selective fat suppression Swap phase-encoding/frequency encoding to reorient artefact away from structures of interest Use VAT, SEMAC, MAVRIC
	Magic Angle	Increased T_2 value of collagenous fibres due to cancellation of dipolar interaction when structure is orientated at the magic angle (55°) with respect to the static magnetic field	Focal signal enhancement of collagen	Increase TE to suppress the collagen signal, reposition patient so structure no longer at magic angle
Inherent physics/sequence parameters	Truncation, Gibbs ringing	Incomplete sampling of echo leading to Fourier truncation effects	Ringing of sharp, high contrast boundaries e.g. edges of brain, spinal cord	Increase matrix size, number of phase-encoding steps
	Phase wrap	Tissues extending outside FOV excited by RF pulse but not correctly phase-encoded Phase contrast: velocity exceeds velocity encoding parameter (VENC)	Tissues 'wrapped around' appearing at opposite side of FOV Phase contrast: sudden change in signal intensity from black to white or vice versa	Increase FOV, re-centre FOV Improved patient positioning, centring Phase oversampling Use surface coil so as not to excite tissues outside FOV Phase contrast MRA: increase VENC
	Cross-talk/ saturation	Imperfect (non-rectangular) slice profile results in overlapping of adjacent slices causing saturation in overlapping region Saturation at overlap of oblique sets of slices	Reduced SNR Reduction in effective TR hence increased apparent T_1 weighting	Interleave slice excitation order If sequential slice order required, increase slice gap to reduce saturation Avoid overlapping oblique sets of slices

Table 8.1 Continued

Type	Artefact	Causes	Appearance	To eliminate/reduce/ workaround
	Chemical shift (imaging)	Difference in resonant frequency between H-nuclei in water and fat molecules (frequency difference = 3.5 ppm) giving rise to chemical shift artefact of the first kind where fat signal is shifted relative to water Chemical shift artefact of the second kind in gradient echo imaging where the TE is such that fat and water signals are out of phase and cancel each other out	Fat shifted in frequency-encoding direction (phase-encoding direction in EPI) giving hyperintense (overlap of fat, water) or hypointense (separation of fat, water) band at fat/water interface. Distance of shift depends on field strength and receiver bandwidth. Direction of shift depends on gradient polarity Chemical shift artefact of the second kind: black boundary where fat and water are out of phase (TE=2.3 ms and every 4.6 ms thereafter at 1.5 T)	Chemical shift artefact of the first kind: Increase bandwidth (=increased gradient amplitude, reduced SNR) Orient low bandwidth direction (frequency-encoding for non-EPI) parallel to fat–water interface Chemical shift artefact of the second kind: use in-phase TEs
	Chemical shift (spectroscopy)	In localised MRS, spatial offset due to chemical shift is applicable to each metabolite and can lead to spatial interference for weakly J-coupled metabolites, e.g. lactate	MRS: changes in expected spectrum due to shifted MRS voxel according to metabolite frequency. Suppression of weakly J-coupled metabolites via spatial interference effect	Use higher bandwidth RF pulses (although not normally definable by general user) Use positive in-phase TEs to avoid suppression of weakly J-coupled metabolites due to spatial interference effect
	Banding artefacts in balanced SSFP	Magnitude of the MR signal varies with the phase evolution of transverse magnetisation, ϕ, occurring between RF pulses from B_0 inhomogeneity, and drops away to zero when $\phi = \pm n180°$ where $n=1,3,5,\ldots$	Bands of signal void where magnetic field inhomogeneities are large, e.g. corners of large FOV	Reduce TR Consider CISS sequence
	B_1 inhomogeneity: dielectric resonance and standing wave effects	RF wavelength in water filled test object becomes more comparable to size of object leading Interference of travelling waves from multiple conducting elements in the quadrature body coil	Non uniformities, localised brightening, localised darkening	Use oil-filled phantoms at 3 T, particularly when assessing uniformity Utilise multiple coils to transmit RF pulses

Table 8.1 Continued

Type	Artefact	Causes	Appearance	To eliminate/reduce/ workaround
	T_2 blurring	Wide spread of echo times in a fast/turbo spin-echo	Image blurring	Reduce echo train length Reduce inter-echo time period by increasing bandwidth Avoid *k*-space ordering schemes that fill from low to high spatial frequencies
	Venetian blind artefact	Use of ramped RF pulses in a hybrid 2D–3D acquisition	Produces gradual variation of signal through image slab with sudden changes at each slab interface giving rise to Venetian blind appearance	Generally not problematic, but some systems offer post-processing to suppress
	Maki artefact	Mistiming of MRA sequence such that presence of contrast agent bolus coincides with the acquisition of the edge rather than centre of *k*-space	Edge enhancement of large blood vessels	Improve timing of MR angiography sequence Consider utilisation of automatic bolus detection techniques
Reconstruction	Concomitant field gradients	Generated by the application of the intended magnetic field gradient as a consequence of Maxwell's equations	Positional offset, ghosting, signal dropout, all increasing as the slice position moves away from isocentre, background phase offsets	Rescan affected slices repositioned at isocentre
	Parallel imaging aliasing	Ill-conditioning – coil sensitivity profiles too similar leading to noise amplification	Localised increased noise, typically in the centre of the phased-array coil	Reduce acceleration factor If 3D sequence, consider distributing acceleration factor over both phase-encoding directions Consider techniques to reduce aliasing, e.g. CAIPIRHINA
	Signal wrap from direct parallel imaging methods	Direct parallel imaging methods are typically unable to deal with signal acquired from outside the selected FOV in the phase-encoding direction	In additional to regular phase wrap at the edges of the FOV, some wrap occurs into the centre of the FOV	Increase FOV to fully encompass the object being imaged Add/increase phase oversampling Switch off parallel imaging Use regenerative (*k*-space) parallel imaging methods
	Coil sensitivity map errors	Errors in applied coil sensitivity map, e.g. as a result of patient movement between acquisition of coil sensitivity map data and imaging data	Irregular localised regions of signal loss	Reacquire normalisation data, try to improve patient compliance

Table 8.1 Continued

Type	Artefact	Causes	Appearance	To eliminate/reduce/ workaround
	Dixon fat–water swaps	Errors in Dixon-based algorithms that aim to separate fat and water signals	Fat and water images, or regions within these images are swapped	As a workaround – always reconstruct both fat and water images Consider more Dixon points if available and practical
	Partial Fourier	Failure of partial Fourier algorithm to reconstruct correctly with the acquired data	May lead to localised signal dropout when echo is shifted outside acquired *k*-space, e.g. as a result of vibration, concomitant fields	Increase fraction of acquired data Cardiac pulsatile effects reduced with zero padding instead of Margosian-based reconstruction
	Intensity scaling	Third-party software may not take into account additional image intensity scaling that is applied differently to images that are combined to form parameter maps	Errors in quantitative maps calculated from original images	Check whether third-party software supports this image intensity scaling Perform calculations on the scanner when available Acquiring the images as part of a single sequence, e.g. multi-echo or dynamic series, then exporting original images may help to avoid variations in image intensity scaling on systems where it is used

ADC, apparent diffusion coefficient; CISS, constructive interference in steady state; CSF, cerebrospinal fluid; EPI, echo planar imaging; FA, fractional anisotropy; FOV, field of view; MRS, MR spectroscopy; SSFP, steady-state free-precession; TE, echo time; TR, repetition time; VENC, velocity sensitivity.

8.7 Acknowledgements

The authors would likc to acknowledge useful discussions with Jonathan Ashmore, Gareth Barker, Craig Buckley, Matthew Clemence, David Collins, David Haire, Dave Higgins, Franklyn Howe, Shaihan Malik, Jonathan Noble and Francesco Padormo, and acknowledge the assistance of MR radiographers at Guy's & St Thomas' NHS Foundation Trust in the collection of example images.

Appendix A References

AAPM (2010) AAPM Report 100. *Acceptance Testing and Quality Assurance Procedures for Magnetic Resonance Imaging Facilities*. American Association of Physicists in Medicine, College Park, MD. Available from http://www.aapm.org/pubs/reports/RPT_100.pdf [accessed on 19/03/2015].

ACR (2005) *ACR Phantom Test Guidance for the ACR MRI Accreditation Program*. American College of Radiology. Available from http://www.acr.org/~/media/ACR/Documents/Accreditation/MRI/LargePhantomGuidance.pdf [accessed on 10/08/2015].

Ai, T., Padua, A., Goerner, F., Nittka, M., Gugala, Z., Jadhav, S., Trelles, M., Johnson, R.F., Lindsey, R.W., Li, X. and Runge, V.M. (2012) SEMAC-VAT and MSVAT-SPACE sequence strategies for metal artifact reduction in 1.5 T magnetic resonance imaging. *Investigative Radiology*, Vol. 47, 267–276.

AnalyzeDirect (2015) Analyze. Available at: http://analyzedirect.com/ [accessed on 4/06/2015].

Ashburner, J. and Friston, K.J. (2000) Voxel-based morphometry – the methods. *Neuroimage*, Vol. 11, 805–821.

Association of Anaesthetists of Great Britain and Ireland (2010) Safety in magnetic resonance units: an update. *Anaesthesia*, Vol. 65, 766–770.

Bailes, D.R., Gilderdale, D.J., Bydder, G.M., Collins, A.G. and Firmin, D.N. (1985) Respiratory ordered phase encoding (ROPE): a method for reducing respiratory motion artefacts in MR imaging. *Journal of Computer Assisted Tomography*, Vol. 9, 835–838.

Baldwin, L.N., Wachowicz, K., Thomas, S.D., Rivest, R. and Fallone, B.G. (2007) Characterization, prediction, and correction of geometric distortion in 3 T MR images. *Medical Physics*, Vol. 34, 388–399.

Bernstein, M.A., Huston, J. and Ward, H.A. (2006) Imaging artifacts at 3.0 T. *Journal of Magnetic Resonance Imaging*, Vol. 24, 735–746.

Bernstein, M.A., King, K.F. and Zhou, X.J. (2004) *Handbook of MRI Pulse Sequences*. Elsevier Academic Press, Burlington, MA.

Bernstein, M.A., Zhou, X.J., Polzin, J.A., King, K.F., Ganin, A., Pelc, N.J. and lover, G.H. (1998) Concomitant gradient terms in phase contrast MR: analysis and correction. *Magnetic Resonance in Medicine*, Vol. 39, 300–308.

Biglands, J.D., Radjenovic, A. and Ridgway, J.P. (2012) Cardiovascular magnetic resonance physics for clinicians: Part II. *Journal of Cardiovascular Magnetic Resonance*, Vol. 14, 66.

Bluml, S., Schad, L.R., Stepanow, B. and Lorenz, W.J. (1993) Spin-lattice relaxation time measurement by means of a TurboFLASH technique. *Magnetic Resonance in Medicine*, Vol. 30, 289–295.

Bodurka, J., Ye, F., Petridou, N., Murphy, K. and Bandettini, P.A. (2007) Mapping the MRI voxel volume in which thermal noise matches physiological noise – implications for fMRI. *Neuroimage*, Vol. 34, 542–549.

Bottomley, P.A., Foster, T.H., Argersinger, R.E. and Pfeiffer, L.M. (1984) A review of normal tissue hydrogen NMR relaxation times and relaxation mechanisms from 1-100 MHz: dependence on tissue type, NMR frequency, temperature, species, excision and age. *Medical Physics*, Vol. 11, 425–448.

Bovee, W.M., Keevil, S.F., Leach, M.O. and Podo, F. (1995) QA in in vivo NMR spectroscopy: II. A protocol for quality assessment. *Magnetic Resonance Imaging*, Vol. 13, 123–129.

Breeuwer, M., Holden, M. and Zylka, W. (2001) Detection and correction of geometric distortion in 3D MR images. *Proceedings of SPIE – the International Society for Optical Engineering*, Vol. 4322, 1110–1120.

Breuer, F.A., Blaimer, M., Heidemann, R.M., Mueller, M.F., Groswold, M.A. and Jakob, P.M. (2005) Controlled aliasing in parallel imaging results in higher acceleration (CAIPIRINHA) for multi-slice imaging. *Magnetic Resonance in Medicine*, Vol. 53, 684–691.

Brookes, J.A., Redpath, T.W., Gilbert, F.J., Murray, A.D. and Staff, R. (1999) Accuracy of T_1 measurement in dynamic contrast-enhanced breast MRI using two- and three-dimensional variable flip angle fast low-angle shot. *Journal of Magnetic Resonance Imaging*, Vol. 9, 163–171.

Bruder, H., Fischer, H., Reinfelder, H.E. and Schmitt, F. (1992) Image reconstruction for echo planar imaging with nonequidistant k-space sampling. *Magnetic Resonance in Medicine*, Vol. 23, 311–323.

Buchenberg, W.B., Dadakova, T., Groebner, J., Bock, M. and Jung, B. (2015) Comparison of two fiber-optical temperature measurement systems in magnetic fields up to 9.4 Tesla. *Magnetic Resonance in Medicine*, Vol. 73, 2047–2051.

Busch, J., Vannesjo, S.J., Barmet, C., Pruessmann, K.P. and Kozerke, S. (2014) Analysis of temperature dependence of background phase errors in phase-contrast cardiovascular magnetic resonance. *Journal of Cardiovascular Magnetic Resonance*, Vol. 16, 97.

Callaghan, M.F., Freund, P., Draganski, B., Anderson, E., Cappelletti, M., Chowdhury, R., Diedrichsen, J., Fitzgerald, T.H., Smittenaar, P., Helms, G., Lutti, A. and Weiskopf, N. (2014) Widespread age-related differences in the human brain microstructure revealed by quantitative magnetic resonance imaging. *Neurobiology of Aging*, Vol. 35, 1862–1872.

Carr, H. (1958) Steady-state free precession in nuclear magnetic resonance. *Physical Review*, Vol. 112, Issue 5, 1693–1701.

Casselman, J.W. and Kuhweide, R. (1993) Constructive interference in steady state 3DFT MR imaging of the inner ear and cerebellopontine angle. *American Journal of Neuroradiology*, Vol. 14, 47–57.

Chang, H. and Fitzpatrick, J.M. (1992) A technique for accurate magnetic resonance imaging in the presence of field inhomogeneities. *IEEE Transactions on Medical Imaging*, Vol. 11, 319–329.

Chen, C.A., Chen, W., Goodman, S.B., Hargreaves, B.A., Koch, K.M., Lu, W., Brau, A.C., Draper, C.E., Delp, S.L. and Gold, G.E. (2011) New MR imaging methods for metallic implants in the knee: Artifact correction and clinical impact. *Journal of Magnetic Resonance Imaging*, Vol. 33, 1121–1127.

Chen, C.C., Wan, Y.L., Wai, Y.Y. and Liu, H.L.J. (2004) Quality assurance of clinical MRI scanners using ACR MRI phantom: preliminary results. *Journal of Digital Imaging*, Vol. 17, 279–284.

Chen, Z., Ma, C.M., Paskalev, K., Li, J., Yang, J., Richardson, T., Palacio, L., Xu, X. and Chen, L. (2006) Investigation of MR image distortion for radiotherapy treatment planning of prostate cancer. *Physics in Medicine and Biology*, Vol. 51, 1393–1403.

Chenevert, T.L., Galbán, C.J., Ivancevic, M.K., Rohrer, S.E., Londy, F.J., Kwee, T.C., Meyer, C.R., Johnson, T.D., Rehemtulla, A. and Ross, B.D. (2011) Diffusion coefficient measurement using a temperature-controlled fluid for quality control in multicenter studies. *Journal of Magnetic Resonance Imaging*, Vol. 34, 983–987.

Chenevert, T.L., Malyarenko, D.I., Newitt, D., Li, X., Jayatilake, M., Tudorica, A., Fedorov, A., Kikinis, R., Liu, T.T., Muzi, M., Oborski, M.J., Laymon, C.M., Li, X., Thomas, Y., Jayashree, K.C., Mountz, J.M., Kinahan, P.E., Rubin, D.L., Fennessy, F., Huang, W., Hylton, N. and Ross, B.D. (2014) Errors in quantitative image analysis due to platform-dependent image scaling. *Translational Oncology*, Vol. 7, 65–71.

Ciarmiello, A., Brunetti, A., Larobina, M., Quarantelli, M., Ziviello, M., Alfano, B. and Salvatore, M. (2001) Assessment of scanner performance and normalization of estimated relaxation rate values. *Magnetic Resonance Imaging*, Vol. 19, 123–128.

Collins, C.M., Liu, W., Schreiber, W., Yang, Q.X. and Smith, M.B. (2005) Central brightening due to constructive interference with, without, and despite dielectric resonance. *Journal of Magnetic Resonance Imaging*, Vol. 21, 192–196.

Colombo, P., Baldassarri, A., Del Corona, M., Mascaro, L. and Strocchi, S. (2004) Multicenter trial for the set-up of a MRI quality assurance programme. *Magnetic Resonance Imaging*, Vol. 22, 93–101.

Constantinides, C.D., Atalar, E. and McVeigh, E.R. (2004) Signal-to-noise measurements in magnitude images from NMR phased arrays. *Magnetic Resonance in Medicine*, Vol. 38, 852–857 (erratum in *Magnetic Resonance in Medicine*, Vol. 52, 219).

Cormack, R.A. (2008) Quality assurance issues for computed tomography-, ultrasound-, and magnetic resonance imaging-guided brachytherapy. *International Journal of Radiation Oncology, Biology, Physics*, Vol. 71, S136–141.

Coward, H.J., O'Gorman, R.L. and McRobbie, D.W. (2007) Assessment of the limiting resolution in MRI. *Proceedings of the International Society for Magnetic Resonance in Medicine*, Vol. 15, 1888.

Crawley, A.P. and Henkelman, R.M. (1987) Errors in T_2 estimation using multislice multiple-echo imaging. *Magnetic Resonance in Medicine*, Vol. 4, 34–47.

de Bazelaire, C.M., Duhamel, G.D., Rofsky, N.M. and Alsop, D.C. (2004) MR imaging relaxation times of abdominal and pelvic tissues measured in vivo at 3.0 T: preliminary results. *Radiology*, Vol. 230, 652–659.

de Certaines, J.D., Henriksen, O., Spisni, A., Cortsen, M. and Ring, P.B. (1993) In vivo measurements of proton relaxation times in human brain, liver and skeletal muscle: a multicenter MRI study. *Magnetic Resonance Imaging*, Vol. 11, 851–856.

De Stefano, N., Filippi, M., Miller, D., Pouwels, PJ., Rovira, A., Gass, A., Enzinger, C., Matthews, P.M. and Arnold, D.L. (2007) Guidelines for using proton MR spectroscopy in multicenter clinical MS studies. *Neurology*, Vol. 69, 1942–1952.

Deibler, A.R., Pollock, J.M., Kraft, R.A., Tan, H., Burdette, J.H. and Maldjian, J.A. (2008) Arterial spin-labeling in routine clinical practice, Part 1: technique and artifacts. *American Journal of Neuroradiology*, Vol. 29, 1228–1234.

Delakis, I., Moore, E.M., Leach, M.O. and De Wilde, J.P. (2004) Developing a quality control protocol for diffusion imaging on a clinical MRI system. *Physics in Medicine and Biology*, Vol. 49, 1409–1422.

Delakis, I., Petala, K. and De Wilde, J.P. (2005) MRI receiver frequency response as a contributor to Nyquist ghosting in echo planar imaging. *Journal of Magnetic Resonance Imaging*, Vol. 22, 324–328.

Deliganis, A.V., Fisher, D.J., Lam, A.M. and Maravilla, K.R. (2001) Cerebrospinal fluid signal intensity increase on FLAIR MR images in patients under general anesthesia: the role of supplemental O2. *Radiology*, Vol. 218, 152–156.

Deoni, S.C.L. (2010) Quantitative relaxometry of the brain. *Topics in Magnetic Resonance Imaging*, Vol. 21, 101–113.

Dietrich, O., Raya, J.G., Reeder, S.B., Ingrisch, M., Reiser, M.F. and Schoenberg, S.O. (2007) Measurement of signal-to-noise ratios in MR images: influence of multichannel coils, parallel imaging, and reconstruction filters. *Journal of Magnetic Resonance Imaging*, Vol. 26, 375–385.

Dietrich, O., Raya, J.G., Reeder, S.B., Ingrisch, M., Reiser, M.F. and Schoenberg, S.O. (2008) Influence of multichannel combination, parallel imaging and other reconstruction techniques on MRI noise characteristics. *Magnetic Resonance Imaging*, Vol. 26, 754–762.

Drost, D.J., Riddle, W.R., Clarke, G.D. AAPM MR Task Group #9 (2002) Proton magnetic resonance spectroscopy in the brain: Report of AAPM MR Task Group #9. *Medical Physics*, Vol. 29, 2177–2197.

Du, G., Lewis, M.M., Styner, M., Shaffer, M.L., Sen, S., Yang, Q.X. and Huang, X. (2011) Combined R_2* and diffusion tenser imaging changes in the substantia nigra in Parkinson disease. *Movement Disorders*, Vol. 26, 1627–1632.

Du, Y.P., Zhou, X.J. and Bernstein, M.A. (2002) Correction of concomitant magnetic field-induced image artifacts in nonaxial echo-planar imaging. *Magnetic Resonance in Medicine*, Vol. 48, 509–515.

Echeverría, J.M.A., Castiella, A. and Emparanza, J.I. (2012) Quantification of iron concentration in the liver by MRI. *Insights into Imaging*, Vol. 3, 173–180.

El-Sharkawy, A.M., Schär, M., Bottomley, P.A. and Atalar, E. (2006) Monitoring and correcting spatio-temporal variations of the MR scanner's static magnetic field. *MAGMA*, Vol. 19, 223–236.

Ernst, T. and Chang, L. (1996) Elimination of artifacts in short echo time 1H MR spectroscopy of the frontal lobe. *Magnetic Resonance in Medicine*, Vol. 36, 462–468.

Ernst, O., Thuret, I., Petit, P., Ameur, F., Loundou, A.D., de Kerviler, E., Izzillo, R., Willig, A.L., Pascal, L., Verlhac, S., Mordon, S., Fenaux, P. and Rose, C. (2013) Iron overload of hematological origin: Validation of a screening procedure for cardiac overload by MRI in routine clinical practice. *Diagnostic and Interventional Imaging*, Vol. 94, 601–608.

European Communities Research Project (1995) Quality assessment in in-vivo NMR spectroscopy: results of a concerted research project for the European Economic Community I–IV. *Magnetic Resonance Imaging*, Vol. 13, 117–158.

European Communities Research Project (COMAC BME II 2.3) (1988) Protocols and test objects for the assessment of MRI equipment. *Magnetic Resonance Imaging*, Vol. 6, 195–199.

Evelhoch, J.L. and Ackerman, J.J.H. (1983) NMR T1 measurements in inhomogeneous B_1 with surface coils. *Journal of Magnetic Resonance (1969)*, Vol. 53, Issue 1, 52–64.

Fahrig, R., Nikolov, H., Fox, A.J. and Holdsworth, D.W. (1999) A three-dimensional cerebrovascular flow phantom. *Medical Physics*, Vol. 26, 1589–1599.

Firbank, M.J., Coulthard, A., Harrison, R.M. and Williams, E.D. (1999) A comparison of two methods for measuring the signal to noise ratio in MR images. *Physics in Medicine and Biology*, Vol. 44, N261–N264.

Firbank, M.J., Harrison, R.M., Williams, E.D. and Coulthard, A. (2000) Quality assurance for MRI: practical experience. *British Journal of Radiology*, Vol. 73, 376–383.

Foerster, B.U., Tomasi, D. and Caparelli, E.C. (2005) Magnetic field shift due to mechanical vibration in functional magnetic resonance imaging. *Magnetic Resonance in Medicine*, Vol. 54, 1261–1267.

Franklin, K.M., Dale, B. and Merkle, E.M. (2008) Improvement in B_1-inhomogeneity artifacts in the abdomen at 3 T MR imaging using a radiofrequency cushion. *Journal of Magnetic Resonance Imaging*, Vol. 27, 1443–1447.

Freed, M., de Zwart, J.A., Hariharan, P., Myers, M.R. and Badano, A. (2011) Development and characterization of a dynamic lesion phantom for the quantitative evaluation of dynamic contrast-enhanced MRI. *Medical Physics*, Vol. 38, 5601–5611.

Friedman, L. and Glover, G.H. (2006) Report on a multicenter fMRI quality assurance protocol. *Journal of Magnetic Resonance Imaging*, Vol. 23, 827–839.

Frigon, C., Jardine, D.S., Weinberger, E., Heckbert, S.R. and Shaw, D.W. (2002) Fraction of inspired oxygen in relation to cerebrospinal fluid hyperintensity on FLAIR MR imaging of the brain in children and young adults undergoing anesthesia. *AJR. American Journal of Roentgenology*, Vol. 179, 791–796.

Galbraith, S.M., Lodge, M.A., Taylor, N.J., Rustin, G.J., Bentzen, S., Stirling, J.J. and Padhani, A.R. (2002) Reproducibility of dynamic contrast-enhanced MRI in human muscle and tumours: comparison of quantitative and semi-quantitative analysis. *NMR in Biomedicine*, Vol. 15, 132–142.

Gallichan, D., Scholz, J., Bartsch, A., Behrens, T.E., Robson, M.D. and Miller, K.L. (2010) Addressing a systematic vibration artifact in diffusion-weighted MRI. *Human Brain Mapping*, Vol. 31, 193–202.

Gatehouse, P.D., Rolf, M.P., Bloch, K.M., Graves, M.J., Kilner, P.J., Firmin, D.N. and Hofman, M.B. (2012) A multi-center inter-manufacturer study of the temporal stability of phase-contrast velocity mapping background offset errors. *Journal of Cardiovascular Magnetic Resonance*, Vol. 14, 72.

Gatehouse, P.D., Rolf, M.P., Graves, M.J., Hofman, M.B., Totman, J., Werner, B., Quest, R.A., Liu, Y., von Spiczak, J., Dieringer, M., Firmin, D.N., van Rossum, A., Lombardi, M., Schwitter, J., Schulz-Menger, J. and Kilner, P.J. (2010) Flow measurement by cardiovascular magnetic resonance: a multi-centre multi-vendor study of background phase offset errors that can compromise the accuracy of derived regurgitant or shunt flow measurements. *Journal of Cardiovascular Magnetic Resonance*, Vol. 12, 5.

Gedamu, E.L., Collins, D.L. and Arnold, D.L. (2008) Automated quality control of brain MR images. *Journal of Magnetic Resonance Imaging*, Vol. 28, 308–319.

Giese, D., Haeberlin, M., Barmet, C., Pruessmann, K.P., Schaeffter, T. and Kozerke, S. (2012) Analysis and correction of background velocity offsets in phase-contrast flow measurements using magnetic field monitoring. *Magnetic Resonance in Medicine*, Vol. 67, 1294–1302.

Gorny, K.R., Hangiandreou, N.J., Hesley, G.K., Gostout, B.S., McGee, K.P. and Felmlee, J.P. (2006) MR guided focused ultrasound: technical acceptance measures for a clinical system. *Physics in Medicine and Biology*, Vol. 51, 3155–3173.

Greve, D.N., Mueller, B.A., Lui, T., Turner, J.A., Voyvodic, J., Yetter, E., Diaz, M., McCarthy, G., Wallace, S., Roach, B.J., Ford, J.M., Mathalon, D.H., Calhoun, V.D., Wible, C.G., Potkin, S.G. and Glover, G (2011) A novel method for quantifying scanner instability in fMRI. *Magnetic Resonance in Medicine*, Vol. 65, 1053–1061.

Griswold, M.A., Kannengiesser, S., Heidemann, R.M., Wang, J. and Jakob, P.M. (2004) Field-of-view limitations in parallel imaging. *Magnetic Resonance in Medicine*, Vol. 52, 1118–1126.

Gu, M., Kim, D.H., Mayer, D., Sullivan, E.V., Pfefferbaum, A. and Spielman, D.M. (2008) Reproducibility study of whole-brain 1H spectroscopic imaging with automated quantification. *Magnetic Resonance in Medicine*, Vol. 60, 542–547.

Gupta, R.K., Ferretti, J., Becker, E. and Weiss, G. (1980) A modified fast inversion-recovery technique for spin-lattice relaxation measurements. *Journal of Magnetic Resonance (1969)*, Vol. 38, 447–452.

Haacke, E.M., Brown, R.W., Thompson, M.R. and Venkatesan, R. (1999) *Magnetic Resonance Imaging: Physical Principles and Sequence Design*. Wiley-Liss, New York.

Haacke, E.M., Xu, Y., Cheng, Y.C. and Reichenbach, J.R. (2004) Susceptibility weighted imaging (SWI). *Magnetic Resonance in Medicine*, Vol. 52, 612–618.

Hariharan, P., Freed, M. and Myers, M.R. (2013) Use of computational fluid dynamics in the design of dynamic contrast enhanced imaging phantoms. *Physics in Medicine and Biology*, Vol. 58, 6369–6391.

Hasan, K.M., Walimuni, I.S., Abid, H., Wolinsky, J.S. and Narayana, P.A. (2012) Multi-modal quantitative MRI investigation of brain tissue neurodegeneration in multiple sclerosis. *Journal of Magnetic Resonance Imaging*, Vol. 35, 1300–1311.

Hattingen, E., Jurcoane, A., Daneshvar, K., Pilatus, U., Mittelbronn, M., Steinbach, J.P. and Bähr, O. (2013) Quantitative T_2 mapping of recurrent glioblastoma under bevacizumab improves monitoring for non-enhancing tumor progression and predicts overall survival. *Neuro-Oncology*, Vol. 15, 1395–1404.

Hattori, K., Ikemoto, Y., Takao, W., Ohno, S., Harimoto, T., Kanazawa, S., Oita, M., Shibuya, K., Kuroda, M. and Kato, H. (2013) Development of MRI phantom equivalent to human tissues for 3.0-T MRI. *Medical Physics*, Vol. 40, 032303.

Haystead, C.M., Dale, B.M. and Merkle, E.M. (2008) N/2 ghosting artifacts: elimination at 3.0-T MR cholangiography with SPACE pulse sequence. *Radiology*, Vol. 246, 589–595.

He, T., Smith, G., Carpenter, J.-P., Mohiaddin, R., Pennell, D. and Firmin, D. (2009) A phantom study of temperature-dependent MRI T_2* measurement. *Journal of Cardiovascular Magnetic Resonance*, Vol. 11, Suppl. 1, 147.

Helms, G. and Piringer, A. (2001) Restoration of motion-related signal loss and line-shape deterioration of proton MR spectra using the residual water as intrinsic reference. *Magnetic Resonance in Medicine*, Vol. 46, 395–400.

Heng, E.L., Vassiliou, V., Gatehouse, P., Greiser, A., Giri, S., Donovan, J., Babu-Narayan, S.V., Gatzoulis, M.A., Pennell, D.J., Prasad, S.K. and Firmin, D. (2014) Temporal stability of gel relaxation-time phantoms for quality control of T_1 and extracellular volume fraction measurements. *Journal of Cardiovascular Magnetic Resonance*, Vol. 16, Suppl. 1, P60.

Hennig, J. (1988) Multiecho imaging sequences with low refocusing flip angles. *Journal of Magnetic Resonance (1969)*, Vol. 78, 397–407.

Henriksen, O., de Certaines, J.D., Spisni, A., Cortsen, M., Muller, R.N. and Ring, P.B. (1993) In vivo dependence of proton relaxation times in human brain, liver and skeletal muscle: a multicenter study. *Magnetic Resonance Imaging*, Vol. 11, 841–850.

Heye, T., Yang, S.R., Bock, M., Brost, S., Weigand, K., Longerich, T., Kauczor, H.U. and Hosch, W. (2012) MR relaxometry of the liver: significant elevation of T_1 relaxation time in patients with liver cirrhosis. *European Radiology*, Vol. 22, 1224–1232.

Hezel, F., Thalhammer, C., Waiczies, S., Schultz-Menger, J. and Nienforf, T. (2012) High spatial resolution and temporally resolved T_2* mapping of normal human myocardium at 7.0 Tesla: an ultrahigh field magnetic resonance feasibility study. *PLoS ONE*, Vol. 7, e52324.

Hines, C.D., Yu, H., Shimakawa, A., McKenzie, C.A., Brittain, J.H. and Reeder, S.B. (2009) T_1 independent, T_2* corrected MRI with accurate spectral modeling for quantification of fat: Validation in a fat-water-SPIO phantom. *Journal of Magnetic Resonance Imaging*, Vol. 30, 1215–1222.

Holland, B.J., Printz, B.F. and Lai, W.W. (2010) Baseline correction of phase-contrast images in congenital cardiovascular magnetic resonance. *Journal of Cardiovascular Magnetic Resonance*, Vol. 12, 11.

Hoult, D.I. and Lauterbur, P.A. (1979) The sensitivity of the zeugmatographic experiment involving human sample. *Journal of Magnetic Resonance*, Vol. 34, 425–433.

Hoult, D.I. and Richards, R.E. (1976) The signal-to-noise ratio of the nuclear magnetic resonance experiment. *Journal of Magnetic Resonance*, Vol. 24, 71–85.

Hunjan Adalsteinsson, E., Kim, D.H., Harsh, G.R., Boyer, A.L., Spielman, D. and Xing, L. (2003) QA of MRS imaging-derived metabolic data. *International Journal of Radiation Oncology, Biology, Physics*, Vol. 57, 1159–1173.

IEC (2007) IEC 62464-1:2007. *Magnetic Resonance Equipment for Medical Imaging – Part 1: Determination of Essential Image Quality Parameters.* International Electrotechnical Commission, Geneva.

Ihalainen, T., Sipilä, O. and Savolainen, S. (2004) MRI quality control: six imagers studied using eleven unified image quality parameters. *European Radiology*, Vol. 14, 1859–1865.

Imran, J., Langevin, F. and Saint-Jalmes, H. (1999) Two-point method for T_1 estimation with optimised gradient-echo sequence. *Magnetic Resonance in Medicine*, Vol. 17, 1347–1356.

Isambert, A., Bonniaud, G., Lavielle, F., Malandain, G. and Lefkopoulos, D. (2008) A phantom study of the accuracy of CT, MR and PET image registrations with a block matching-based algorithm. *Cancer Radiothérapie*, Vol. 12, 800–808.

Jackson, A., Buckley, D.L. and Parker, G.J.M. (eds) (2005) *Dynamic Contrast-enhanced Magnetic Resonance Imaging in Oncology*. Springer, Berlin, Heidelberg.

Jackson, A., Jayson, G.C., Li, K.L., Zhu, X.P., Checkley, D.R., Tessier, J.J. and Waterton, J.C. (2003) Reproducibility of quantitative dynamic contrast-enhanced MRI in newly presenting glioma. *British Journal of Radiology*, Vol. 76, 153–162.

Jezzard, P. (2009) The physical basis of spatial distortions in magnetic resonance images. In: *Handbook of Medical Image Processing and Analysis*, 2nd edition, edited by I.N. Bankman. Elsevier, Amsterdam, 499–514.

Jezzard, P., Matthews, P.M. and Smith, S.M. (eds) (2001) *Functional Magnetic Resonance Imaging: An Introduction to Methods*. Oxford University Press, Oxford.

Judy, P.F. (1976) The line spread function and modulation transfer function of a computed tomography scanner. *Medical Physics*, Vol. 3, 233–236.

Kaldoudi, E. and Williams, S.C.R. (1993) Relaxation time measurements in NMR imaging. Part I: Longitudinal relaxation time. *Concepts in Magnetic Resonance*, Vol. 5, 217–242.

Kaufman, L., Kramer, D.M., Crooks, L.E. and Ortendahl, D.A. (1989) Measuring signal-to-noise ratios in MR imaging. *Radiology*, Vol. 173, 265–267.

Kellman, P., Wilson, J.R., Xue, H., Ugander, M. and Arai, A.E. (2012) Extracellular volume fraction mapping in the myocardium, part 1: evaluation of an automated method. *Journal of Cardiovascular Magnetic Resonance*, Vol. 14, 63.

Khan, S. and de Wilde, J. (2003) Design of a dynamic cardiac MR phantom for the evaluation of cardiac MR systems. *Computer Assisted Radiology and Surgery*, Vol. 1256, 1165–1170.

Khoo, V.S., Dearnaley, D.P., Finnigan, D.J., Padhani, A., Tanner, S.F. and Leach, M.O. (1997) Magnetic resonance imaging (MRI): considerations and applications in radiotherapy treatment planning. *Radiotherapy and Oncology*, Vol. 42, 1–15.

Kingsley, P.B. (1999a) Signal intensities and T_1 calculations in multiple-echo sequences with imperfect pulses. *Concepts in Magnetic Resonance*, Vol. 11, 29–49.

Kingsley, P.B. (1999b) Methods of measuring spin-lattice (T_1) relaxation times: An annotated bibliography. *Concepts in Magnetic Resonance*, Vol. 11, 243–276.

Kingsley, P.B., Ogg, R.J., Reddick, W.E. and Steen, R.G. (1998) Correction of errors caused by imperfect inversion pulses in MR imaging measurement of T_1 relaxation times. *Magnetic Resonance Imaging*, Vol. 16, 1049–1055.

Klose, U. (1990) In vivo proton spectroscopy in presence of eddy currents. *Magnetic Resonance in Medicine*, Vol. 14, 26–30.

Koch, K.M., Lorbiecki, J.E., Hinks, R.S. and King, K.F. (2009) A multispectral three-dimensional acquisition technique for imaging near metal implants. *Magnetic Resonance in Medicine*, Vol. 61, 381–390.

Koh, D.M., Blackledge, M., Padhani, A.R., Takahara, T., Kwee, T.C., Leach, M.O. and Collins, D.J. (2012) Whole-body diffusion-weighted MRI: Tips, tricks, and pitfalls. *AJR. American Journal of Roentgenology*, Vol. 199, 252–262.

Koller, C.J., Etough, J.P., Mountford, P.J. and Frain, G. (2006) A survey of MRI quality assurance programmes. *British Journal of Radiology*, Vol. 79, 592–596.

Kondur, A.K., Li, T., Vaitkevicius, P. and Afonso, L. (2009) Quantification of myocardial iron overload by cardiovascular magnetic resonance imaging T_2* and review of the literature. *Clinical Cardiology*, Vol. 32, E55–59.

Kreis, R. (2004) Issues of spectral quality in clinical 1H-magnetic resonance spectroscopy and a gallery of artifacts. *NMR in Biomedicine*, Vol. 17, 361–381.

Kristensen, B.H., Laursen, F.J., Logager, V., Geertsen, P.F. and Krarup-Hansen, A. (2008) Dosimetric and geometric evaluation of an open low-field magnetic resonance simulator for radiotherapy treatment planning of brain tumours. *Radiotherapy and Oncology*, Vol. 87, 100–109.

Kuhl, C.K., Kooijman, H., Gieseke, J. and Schild, H.H. (2007) Effect of B_1 inhomogeneity on breast MR imaging at 3.0 T. *Radiology*, Vol. 244, 929–930.

Lange, T., Dydak, U., Roberts, T.P., Rowley, H.A., Bjeljac, M. and Boesiger, P. (2006) Pitfalls in lactate measurements at 3 T. *American Journal of Neuroradiology*, Vol. 27, 895–901.

Lange, T., Zaitsev, M. and Buechert, M. (2011) Correction of frequency drifts induced by gradient heating in 1H spectra using interleaved reference spectroscopy. *Journal of Magnetic Resonance Imaging*, Vol. 33, 748–754.

Lankhaar, J.W., Hofman, M.B., Marcus, J.T., Zwanenburg, J.J., Faes, T.J. and Vonk-Noordegraaf, A. (2005) Correction of phase offset errors in main pulmonary artery flow quantification. *Journal of Magnetic Resonance Imaging*, Vol. 22, 73–79.

Larsson, H.B., Rosenbaum, S. and Fritz-Hansen, T. (2001) Quantification of the effect of water exchange in dynamic contrast MRI perfusion measurements in the brain and heart. *Magnetic Resonance in Medicine*, Vol. 46, 272–281.

Laubach, H.J., Jakob, P.M., Loevblad, K.O., Baird, A.E., Bovo, M.P., Edelman, R.R. and Warach, S. (1998) A phantom for diffusion-weighted imaging of acute stroke. *Journal of Magnetic Resonance Imaging*, Vol. 8, 1349–1354.

Lavdas, I., Behan, K., Papadaki, A., McRobbie, D.W. and Aboagye, E.O. (2013) A phantom for diffusion weighted magnetic resonance imaging (DW-MRI). *Journal of Magnetic Resonance Imaging*, Vol. 38, 173–179.

Leach, M.O., Collins, D.J., Keevil, S., Rowland, I., Smith, M.A., Henriksen, O., Bovée, W.M. and Podo, F. (1995) QA in in vivo NMR spectroscopy: III. Clinical test objects: design, construction and solutions. *Magnetic Resonance Imaging*, Vol. 13, 131–137.

Lebel, R.M. and Wilman, A.H. (2010) Transverse relaxometry with stimulated echo compensation. *Magnetic Resonance in Medicine*, Vol. 64, 1005–1014.

Lee, V.S., Martin, D.J., Krinsky, G.A. and Rofsky, N.M. (2000) Gadolinium-enhanced MR angiography: artifacts and pitfalls. *AJR. American Journal of Roentgenology*, Vol. 175, 197–205.

Lerski, D. (2003) *Eurospin II Magnetic Resonance Quality Assessment Test Objects: Instructions for Use*. Diagnostic Sonar Ltd, Livingston, UK.

Lerski, R.A. and de Certaines, J.D. (1993) Performance assessment and quality control in MRI by Eurospin test objects and protocols. *Magnetic Resonance Imaging*, Vol. 11, 817–833.

Lerski, R.A., McRobbie, D.W., Straughan, K., Walker, P.M., de Certaines, J.D. and Bernard, A.M. (1988) Multi-center trial with protocols and prototype test objects for the assessment of MRI equipment. EEC Concerted Research Project. *Magnetic Resonance Imaging*, Vol. 6, 201–214.

Lerski, R.A., de Wilde, J., Boyce, D. and Ridgway, J. (1998) *Quality Control in Magnetic Resonance Imaging*. IPEM Report 80. Institute of Physics and Engineering in Medicine, York, UK.

Li, S., Sun, F., Jin, Z.Y., Xue, H.D. and Li, M.L. (2007) Whole-body diffusion-weighted imaging: technical improvement and preliminary results. *Journal of Magnetic Resonance Imaging*, Vol. 26, 1139–1144.

Liang, Z.-P., Boada, F.E., Constable, R.T., Haake, M.E., Lauterbur, P.C. and Smith, M.R. (1992) Constrained reconstruction methods in MR imaging. *Reviews of Magnetic Resonance in Medicine*, Vol. 4, 67–185.

Lorrain, P. and Corson, D. (1970) *Electromagnetic Fields and Waves*, 2nd edition. WH Freeman, New York.

Lotz, J., Meier, C., Leppert, A. and Galanski, M. (2002) Cardiovascular flow measurement with phase-contrast MR imaging: basic facts and implementation. *Radiographics*, Vol. 22, 651–671.

Ma, J. and Zhou, X. (2002) *Fast Spin Echo Phase Correction for MRI System*. US 6369568 B1. US Patent Office, Alexandria, Virginia.

MagNET (2006) *MagNET Test Objects: Instructions for Use*. Version 4. MagNET, London.

Majumdar, S., Orphanoudakis, S.C., Gmitro, A., O'Donnell, M. and Gore, J.C. (1986) Errors in the measurements of T_2 using multiple-echo MRI techniques. I. Effects of radiofrequency pulse imperfections. *Magnetic Resonance in Medicine*, Vol. 3, 397–417.

Maki, J.H., Prince, M.R., Londy, F.J. and Chenevert, T.L. (1996) The effects of time varying intravascular signal intensity and k-space acquisition order on three-dimensional MR angiography image quality. *Journal of Magnetic Resonance Imaging*, Vol. 6, 642–651.

Marchand, B., Douek, P.C., Robert, P., Corot, C., Roux, J.P., Adeleine, P., Hernandez-Hoyos, M., Cremillieux, Y., Orkisz, M. and Canet, E. (2002) Standardized MR protocol for the evaluation of MRA sequences and/or contrast agents effects in high-degree arterial stenosis analysis. *MAGMA*, Vol. 14, 259–267.

Marjanska, M., Waks, M., Snyder, C.J. and Vaughan, J.T. (2008) Multinuclear NMR investigation of probe construction materials at 9.4 T. *Magnetic Resonance in Medicine*, Vol. 59, 936–938.

Markl, M., Bammer, R., Alley, M.T., Elkins, C.J., Draney, M.T., Barnett, A., Moseley, M.E., Glover, G.H. and Pelc, N.J. (2003) Generalized reconstruction of phase contrast MRI: analysis and correction of the effect of gradient field distortions. *Magnetic Resonance in Medicine*, Vol. 50, 791–801.

MathWorks (2015) *MATLAB*. Available at: http://uk.mathworks.com/products/matlab/ [accessed on 4/06/2015].

McCann, A.J., Workman, A. and McGrath, C. (2013) A quick and robust method for measurement of signal-to-noise ratio in MRI. *Physics in Medicine and Biology*, Vol. 58, 3775–3790.

McGibney, G., Smith, M.R., Nichols, S.T. and Crawley, A. (1993) Quantitative evaluation of several partial Fourier reconstruction algorithms used in MRI. *Magnetic Resonance in Medicine*, Vol. 30, 51–59.

McRobbie, D.W. (1996) Short Communication: The absolute signal-to-noise ratio in MRI acceptance testing. *British Journal of Radiology*, Vol. 69, 1045–1048.

McRobbie, D.W. (1997) A 3-D volumetric test object for geometry evaluation in MRI. *Medical Physics*, Vol. 24, 743–750.

McRobbie, D.W. (2010) Magnetic resonance imaging and EM field exposure. In: *Guidance on Electromagnetic Fields and Electromagnetic Compatibility,* edited by M. Robinson, L. Grant and J. Reeves. IPEM Report 98. Institute of Physics and Engineering in Medicine, York, UK, 74–86.

McRobbie, D.W., Lerski, R.A., Straughan, K., Quilter, P. and Orr, J.S. (1986) Investigation of slice characteristics in nuclear magnetic resonance imaging. *Physics in Medicine and Biology*, Vol. 31, 613–626.

McRobbie, D.W., Moore, E.M., Graves, M.J. and Prince, M.R. (2007) *MRI: from Picture to Proton,* 2nd edition. Cambridge University Press, Cambridge.

McRobbie, D.W. and Quest, R.A. (2002) Effectiveness and relevance of MR acceptance testing: Results of an eight year audit. *British Journal of Radiology*, Vol. 75, 523–531.

Meier, C., Zwanger, M., Feiweier, T. and Porter, D. (2008) Concomitant field terms for asymmetric gradient coils: consequences for diffusion, flow, and echo-planar imaging. *Magnetic Resonance in Medicine*, Vol. 60, 128–134.

Meng, Y. and Lei, H. (2008) A single-scan T_2* mapping method based on two gradient-echo images with compensation for macroscopic field inhomogeneity. *Magnetic Resonance in Medicine*, Vol. 60, 1388–1395.

Messroghli, D. (2015) *MRmap*. Available at http://sourceforge.net/projects/mrmap/ [accessed on 4/06/2015].

MHRA (2014) *Safety Guidelines for Magnetic Resonance Imaging Equipment in Clinical Use.* MHRA DB2007. Medicines and Healthcare products Regulatory Agency, London.

Miller, T.A., Landes, A.B. and Moran, A.M. (2009) Improved accuracy in flow mapping of congenital heart disease using stationary phantom technique. *Journal of Cardiovascular Magnetic Resonance*, Vol. 11, 52.

Mitchell, M.D., Kundel, H.L., Axel, L. and Joseph, P.M. (1986) Agarose as a tissue equivalent phantom material for NMR imaging. *Magnetic Resonance Imaging*, Vol. 4, 263–266.

Moorhead, T.W., Gountouna, V.E., Job, D.E., McIntosh, A.M., Romaniuk, L., Lymer, G.K., Whalley, H.C., Waiter, G.D., Brennan, D., Ahearn, T.S., Cavanagh, J., Condon, B., Steele, J.D., Wardlaw, J.M. and Lawrie, S.M. (2009) Prospective multi-centre Voxel Based Morphometry study employing scanner specific segmentations: procedure development using CaliBrain structural MRI data. *BMC Medical Imaging*, Vol. 9, 8.

Mugler, J.P., Bao, S., Mulkern, R.V., Guttmann, C.R., Robertson, R.L., Jolesz, F.A. and Brookeman, J.R. (2000) Optimized single-slab three-dimensional spin-echo MR imaging of the brain. *Radiology*, Vol. 216(3), 891–899.

Mulkern, R.V., Forbes, P., Dewey, K., Osganian, S., Clark, M., Wong, S., Ramamurthy, U., Kun, L. and Poussaint, T.Y. (2008) Establishment and results of a magnetic resonance quality assurance program for the pediatric brain tumor consortium. *Academic Radiology*, Vol. 15, 1099–1110.

Nelson, T.R. and Tung, S.M. (1987) Temperature dependence of proton relaxation times in vitro. *Magnetic Resonance Imaging*, Vol. 5, 189–199.

NEMA (2008a) *NEMA Standards Publication MS 1-2008, Determination of Signal-to-Noise Ratio (SNR) in Diagnostic Magnetic Resonance Imaging*. National Electrical Manufacturers Association, Rosslyn, VA, USA. Available from http://www.nema.org/Standards/Pages/Determination-of-Signal-to-Noise-Ratio-in-Diagnostic-Magnetic-Resonance-Imaging.aspx [accessed on 10/08/2015].

NEMA (2008b) *NEMA Standards Publication MS 2-2008, Determination of Two-Dimensional Geometric Distortion in Diagnostic Magnetic Resonance Images.* National Electrical Manufacturers Association, Rosslyn, VA, USA. Available from http://www.nema.org/Standards/Pages/Determination-of-Two-Dimensional-Geometric-Distortion-in-Diagnostic-Magnetic-Resonance-Images.aspx [accessed on 10/08/2015].

NEMA (2008c) *NEMA Standards Publication MS 3-2008, Determination of Image Uniformity in Diagnostic Magnetic Resonance Images.* National Electrical Manufacturers Association, Rosslyn, VA, USA. Available from http://www.nema.org/Standards/Pages/Determination-of-Image-Uniformity-in-Diagnostic-Magnetic-Resonance-Images.aspx [accessed on 10/08/2015].

NEMA (2008d) *NEMA Standards Publication MS 6-2008, Determination of Signal-to-Noise Ratio and Image Uniformity for Single-Channel Non-Volume Coils in Diagnostic MR Imaging.* National Electrical Manufacturers Association, Rosslyn, VA, USA. Available from http://www.nema.org/Standards/Pages/Determination-of-Signal-to-Noise-Ratio-and-Image-Uniformity-for-Single-Channel-Non-Volume-Coils-in-Diagnostic-Magnetic.aspx [accessed on 10/08/2015].

NEMA (2008e) *NEMA Standards Publication MS 9-2008, Characterisation of Phased Array Coils for Diagnostic Magnetic Resonance Images.* National Electrical Manufacturers Association, Rosslyn, VA, USA. Available from http://www.nema.org/Standards/Pages/Characterisation-of-Phased-Array-Coils-for-Diagnostic-Magnetic-Resonance-Images.aspx [accessed on 10/08/2015].

NEMA (2010a) *NEMA Standards Publication MS 5-2010, Determination of Slice Thickness in Diagnostic Magnetic Resonance Imaging.* National Electrical Manufacturers Association, Rosslyn, VA, USA. Available from http://www.nema.org/Standards/Pages/Determination-of-Slice-Thickness-in-Diagnostic-Magnetic-Resonance-Imaging.aspx [accessed on 10/08/2015].

NEMA (2010b) *NEMA Standards Publication MS 12-2010, Quantification and Mapping of Geometric Distortion for Special Applications.* National Electrical Manufacturers Association, Rosslyn, VA, USA. Available from http://www.nema.org/Standards/Pages/Quantification-and-Mapping-of-Geometric-Distortion-for-Special-Applications.aspx [accessed on 10/08/2015].

NEMA (2011) *Digital Imaging and Communications in Medicine (DICOM) Part 3: Information Object Definitions.* PS 3.3-2011 Section C.8.3.1 MR Image Module. National Electrical Manufacturers Association, Rosslyn, VA, USA. Available from http://medical.nema.org/Dicom/2011/11_03pu.pdf [accessed on 10/08/2015].

NEMA (2016) *Information Object Definitions.* PS 3.3-2016b Section 8.3.1 MR Image Module. Digital Imaging and Communications in Medicine (DICOM) Standard, National Electrical Manufacturers Association, Rosslyn, VA, USA. Available from http://dicom.nema.org/standard.html

NIH (2015) *ImageJ.* National Institutes of Health, Bethesda, MD, USA. Available at: http://imagej.nih.gov/ij/ [accessed on 10/08/2015].

Oakland, J.S. (1996) *Statistical Process Control: A Really Practical Guide,* 3rd edition, Butterworth-Heinemann, Oxford, Boston.

Ocali, O. and Atalar, E. (1998) Ultimate intrinsic signal-to-noise ratio in MRI. *Magnetic Resonance in Medicine*, Vol. 39, 462–473.

Ohliger, M.A., Grant, A.K. and Sodickson, D.K. (2003) Ultimate intrinsic signal-to-noise ratio for parallel MRI: electromagnetic field considerations. *Magnetic Resonance in Medicine*, Vol. 50, 1018–1030.

Olsrud, J., Nilsson, A., Mannfolk, P., Waites, A. and Ståhlberg, F. (2007) A two-compartment gel phantom for optimization and quality assurance in clinical BOLD fMRI. *Magnetic Resonance Imaging*, Vol. 26, 279–286.

Orr, T.N., Paliy, I., Winter, J., Campbell, G., Thompson, R.T. and Gelman, N. (2007) Development of a composite material phantom mimicking the magnetic resonance parameters of the neonatal brain at 3.0 Tesla. *Investigative Radiology*, Vol. 42, 739–746.

Parker, D.L., Yuan, C. and Blatter, D.D. (1991) MR angiography by multiple thin slab 3D acquisition. *Magnetic Resonance in Medicine*, Vol. 17, 434–451.

Parker, G.J.M., Barker, G.J. and Tofts, P.S. (2001) Accurate multislice gradient echo T_1 measurement in the presence of non-ideal RF pulse shape and RF field nonuniformity. *Magnetic Resonance in Medicine*, Vol. 45, 838–845.

Pell, G.S., Briellmann, R.S., Waites, A.B., Abbott, D.F., Lewis, D.P. and Jackson, G.D. (2006) Optimized clinical T_2 relaxometry with a standard CPMG sequence. *Journal of Magnetic Resonance Imaging*, Vol. 23, 248–252.

Pennell, D.J. (2005) T_2* magnetic resonance and myocardial iron in thalassemia. *Annals of the New York Academy of Sciences*, Vol. 1054, 373–378.

Pfefferbaum, A., Sullivan, E.V., Hedehus, M., Moseley, M. and Lim, K.O. (1999) Brain gray and white matter transverse relaxation time in schizophrenia. *Psychiatry Research*, Vol. 11, 425–448.

Pierpaoli, C., Jezzard, P., Baser, P.J., Barnett, A. and Di Chiro, G. (1996) Diffusion tensor MR imaging of the human brain. *Radiology*, Vol. 201, 637–648.

Pipe, J.G. (1999) Motion correction with PROPELLER MRI: application to head motion and free-breathing cardiac imaging. *Magnetic Resonance in Medicine*, Vol. 42, 963–969.

Poetker, D.M., Jursinic, P.A., Runge-Samuelson, C.L. and Wackym, P.A. (2005) Distortion of magnetic resonance images used in gamma knife radiosurgery treatment planning: implications for acoustic neuroma outcomes. *Otology & Neurotology*, Vol. 26, 1220–1228.

Polders, D.L., Leemans, A., Hendrikse, J., Donahue, M.J., Luijten, P.R. and Hoogduin, J.M. (2011) Signal to noise ratio and uncertainty in diffusion tensor imaging at 1.5, 3.0 and 7.0 tesla. *Journal of Magnetic Resonance Imaging*, Vol. 33, 1456–1463.

Posse, S., Cuenod, C.A. and Le Bihan, D. (1993) Motion artifact compensation in 1H spectroscopic imaging by signal tracking. *Journal of Magnetic Resonance. Series B*, Vol. 102, 222–227.

Preibisch, C., Volz, S., Anti, S. and Deichmann, R. (2008) Exponential excitation pulses for improved water content mapping in the presence of background gradients. *Magnetic Resonance in Medicine*, Vol. 60, 908–916.

Price, R.R., Axel, L., Morgan, T., Newman, R., Perman, W., Schneiders, N., Selikson, M., Wood, M. and Thomas, S.R. (1990) Quality assurance methods and phantoms for magnetic resonance imaging: Report of AAPM Nuclear Magnetic Resonance Task Group No. 1a. *Medical Physics*, Vol. 17, 287–295.

Pullens, P., Roebroeck, A. and Goebel, R. (2010) Ground truth hardware phantoms for validation of diffusion-weighted MRI applications. *Journal of Magnetic Resonance Imaging*, Vol. 23, 482–488.

Purdey, D., Cadena, G. and Laub, G. (1992) The design of variable tip angle slab selection (TONE) pulses for improved 3-D MR angiography. In: *Proceedings of the Eleventh Meeting of the Society of Magnetic Resonance in Medicine*, 882.

Rajan, S., Herbertson, L., Bernardo, M. and Choyke, P. (2014) A dialyzer-based flow system for validating dynamic contrast enhanced MR image acquisition. *Magnetic Resonance in Medicine*, Vol. 72, 41–48.

Redpath, T.W. (1998) Signal-to-noise ratio in MRI. *British Journal of Radiology*, Vol. 71, 704–707.

Redpath, T.W. and Wiggins, C.J. (2000) Estimating achievable signal-to-noise ratios of MRI transmit-receive coils from radiofrequency power measurements: applications in quality control. *Physics in Medicine and Biology*, Vol. 45, 217–227.

Reese, T.G., Heid, O., Weisskoff, R.M. and Wedeen, V.J. (2002) Reduction of eddy-current-induced distortion in diffusion MRI using a twice-refocused spin echo. *Magnetic Resonance in Medicine*, Vol. 49, 177–182.

Reischauer, C., Staempfli, P., Jaermann, T. and Boesiger, P. (2009) Construction of a temperature-controlled diffusion phantom for quality control of diffusion measurements. *Journal of Magnetic Resonance Imaging*, Vol. 29, 692–698.

Rigsby, C.K., Hilpipre, N., McNeal, G.R., Zhang, G., Boylan, E.E., Popescu, A.R., Choi, G., Greiser, A. and Deng, J. (2013) Analysis of an automated background correction method for cardiovascular MR phase contrast imaging in children and young adults. *Pediatric Radiology*, Vol. 29, 113–120.

Robson, M.D. and Porter, D.A. (2005) Reconstruction as a source of artifact in non-gated single-shot diffusion-weighted EPI. *Magnetic Resonance Imaging*, Vol. 23, 899–905.

Roebuck, J.R., Haker, S.J., Mitsouras, D., Rybicki, F.J., Tempany, C.M. and Mulkern, R.V. (2009) Carr-Purcell-Meiboom-Gill imaging of prostate cancer: quantitative T_2 values for cancer discrimination. *Magnetic Resonance Imaging*, Vol. 27, 497–502.

Roemer, P.B., Edelstein, W.A., Hayes, C.E., Souza, S.P. and Mueller, O.M. (1990) The NMR phased array. *Magnetic Resonance in Medicine*, Vol. 16(2), 192–225.

Rolf, M.P., Hofman, M.B., Gatehouse, P.D., Markenroth-Bloch, K., Heymans, M.W., Ebbers, T., Graves, M.J., Totman, J.J., Werner, B., van Rossum, A.C., Kilner, P.J. and Heethaar, R.M. (2011) Sequence optimization to reduce velocity offsets in cardiovascular magnetic resonance volume flow quantification – a multi-vendor study. *Journal of Cardiovascular Magnetic Resonance*, Vol. 13, 18.

Scheenen, T.W., Klomp, D.W., Wijnen, J.P. and Heerschap, A. (2008) Short echo time 1H-MRSI of the human brain at 3 T with minimal chemical shift displacement errors using adiabatic refocusing pulses. *Magnetic Resonance in Medicine*, Vol. 59, 1–6.

Sedlacik, J., Boelmans, K., Löbel, U., Holst, B., Siemonsen, S. and Fiehler, J. (2014) Reversible, irreversible and effective transverse relaxation rates in normal aging brain at 3 T. *NeuroImage*, Vol. 84, 1032–1041.

Seitz, J., Strotzer, M., Völk, M., Held, P., Djavidani, B., Nitz, W.R. and Feuerbach, S. (2000) Reduction of motion artifacts in magnetic resonance imaging of the neck and cervical spine by long-term averaging. *Investigative Radiology*, Vol. 35, 380–384.

Sener, R.N. (2001) Diffusion MRI: apparent diffusion coefficient (ADC) values in the normal brain and a classification of brain disorders based on ADC values. *Computerized Medical Imaging and Graphics*, Vol. 25, 299–326.

Shanmugalingam, S., Thornton, J.S., Iwata, O., Bainbridge, A., O'Brien, F.E., Priest, A.N., Ordidge, R.J., Cady, E.B., Wyatt, J.S. and Robertson, N.J. (2006) Comparative prognostic utilities of early quantitative magnetic resonance imaging spin-spin relaxometry and proton magnetic resonance spectroscopy in neonatal encephalopathy. *Pediatrics*, Vol. 118, 1467–1477.

Shin, W., Gu, H. and Yang, Y. (2009) Fast high-resolution T_1 mapping using inversion-recovery Look-Locker echo-planar imaging at steady state: optimization for accuracy and reliability. *Magnetic Resonance in Medicine*, Vol. 61, 899–906.

Shrager, R.I., Weiss, G.H. and Spencer, R.G. (1998) Optimal time spacings for T_2 measurements: monoexponential and biexponential systems. *NMR in Biomedicine*, Vol. 11, 297–305.

Simmons, A., Moore, E. and Williams, S.C.R. (1999) Quality control for functional magnetic resonance imaging using automated data analysis and Shewhart charting. *Magnetic Resonance in Medicine*, Vol. 41, 1274–1278.

Sled, J.G. and Nossin-Manor, R. (2013) Quantitative MRI for studying neonatal brain development. *Neuroradiology*, Vol. 55, 97–104.

Smith, S.M., Jenkinson, M., Woolrich, M.W., Beckmann, C.F., Behrens, T.E., Johansen-Berg, H., Bannister, P.R., De Luca, M., Drobnjak, I., Flitney, D.E., Niazy, R.K., Saunders, J., Vickers, J., Zhang, Y., De Stefano, N., Brady, J.M. and Matthews, P.M. (2004) Advances in functional and structural MR image analysis and implementation as FSL. *NeuroImage*, Vol. 23, Suppl 1, S208–219.

Sourbron, S.P. and Buckley, D.L. (2013) Classic models for dynamic contrast-enhanced MRI. *NMR in Biomedicine*, Vol. 28, 1004–1027.

St. Pierre, T.G., Clark, P.R. and Chua-Anusorn, W. (2005) Measurement and mapping of liver iron concentrations using magnetic resonance imaging. *Annals of the New York Academy of Sciences*, Vol. 1054, 379–385.

Stadler, A., Schima, W., Ba-Ssalamah, A., Kettenbach, J. and Eisenhuber, E. (2007). Artifacts in body MR imaging: their appearance and how to eliminate them. *European Radiology*, Vol. 17, 1242–1255.

Steckner, M.C. (2010). A simple method for estimating the noise level in a signal region of an MR image. *Medical Physics*, Vol. 37, 5072–5079.

Steckner, M.C., Lui, B. and Ying, L. (2009). A new single acquisition, two-image difference method for determining MR image SNR. *Medical Physics*, Vol. 36, 662–671.

Steinhoff, S., Zaitsev, M., Zilles, K. and Shah, N.J. (2001). Fast T_1 mapping with volume coverage. *Magnetic Resonance in Medicine*, Vol. 46, 131–140.

Stonnington, C.M., Tan, G., Klöppel, S., Chu, C., Draganski, B., Jack, C.R. Jr, Chen, K., Ashburner, J. and Frackowiak, R.S. (2008). Interpreting scan data acquired from multiple scanners: a study with Alzheimer's disease. *Neuroimage*, Vol. 39, 1180–1185.

Suckling, J., Barnes, A., Job, D., Brenan, D., Lymer, K., Dazzan, P., Marques, T.R., MacKay, C., McKie, S., Williams, S.R., Williams, S.C., Lawrie, S. and Deakin, B. (2010). Power calculations for multicenter imaging studies controlled by the false discovery rate. *Human Brain Mapping*, Vol. 31, 1183–1189.

Sun, J., Barnes, M., Dowling, J., Menk, F., Stanwell, P. and Greer, P.B. (2015). An open source automatic quality assurance (OSAQA) tool for the ACR MRI phantom. *Australasian Physical & Engineering Sciences in Medicine*, Vol. 38, 39–46.

Systat Software Inc. (2015). *SigmaPlot*. Systat Software Inc., San Jose, CA. Available at: http://www.sigmaplot.co.uk/ [accessed on 4/06/2015].

Tang, A.M., Kacher, D.F., Lam, E.Y., Wong, K.K., Jolesz, F.A. and Yang, E.S. (2008). Simultaneous ultrasound and MRI system for breast biopsy: compatibility assessment and demonstration in a dual modality phantom. *IEEE Transactions on Medical Imaging*, Vol. 27, 247–254.

Tofts, P. (1990). Standing waves in uniform water phantoms. *Journal of Magnetic Resonance. Series B*, Vol. 17, 287–295.

Tofts, P.S. (2003a). The measurement process: MR data collection and image analysis. In: *Quantitative MRI of the Brain: Measuring Changes Caused by Disease,* edited by P. Tofts. Wiley, Chichester, UK, 17–54.

Tofts, P.S. (2003b) Quality assurance, accuracy, precision and phantoms. In: *Quantitative MRI of the Brain: Measuring Changes Caused by Disease*, edited by P. Tofts. Wiley, Chichester, UK.

Tofts, P.S. and Collins, D.J. (2011) Multicentre imaging measurements for oncology and in the brain. *British Journal of Radiology*, Vol. 84, S213–226.

Tofts, P.S. and Parker, G.J.M. (2013) DCE-MRI: acquisition and analysis techniques. In: *Clinical Perfusion MRI Techniques and Applications*, edited by P.B. Barker, X.G. Golay and G. Zaharchuk. Cambridge University Press, Cambridge, UK.

Tofts, P.S., Jackson, J.S., Tozer, D.J., Cercignani, M., Keir, G., MacManus, D.G., Ridgway, G.R., Ridha, B.H., Schmierer, K., Siddique, D., Thornton, J.S., Wroe, S.J. and Fox, N.C. (2008) Imaging cadavers: cold FLAIR and noninvasive brain thermometry using CSF diffusion. *Magnetic Resonance in Medicine*, Vol. 59, 190–195.

Tofts, P.S., Lloyd, D., Clark, C.A., Barker, G.J., Parker, G.J.M., McConville, P., Baldock, C. and Pop, J.M. (2000) Test liquids for quantitative MRI measurements of self-diffusion coefficient in vivo. *Magnetic Resonance in Medicine*, Vol. 43, 368–374.

van der Graaf, M., Julià-Sapé, M., Howe, F.A., Ziegler, A., Majós, C., Moreno-Torres, A., Rijpkema, M., Acosta, D., Opstad, K.S., van der Meulen, Y.M., Arús, C. and Heerschap, A. (2008) MRS quality assessment in a multi-centre study on MRS-based classification of brain tumours. *NMR in Biomedicine*, Vol. 21, 148–158.

Velthuizen, R.P., Heine, J.J., Cantor, A.B., Lin, H., Fletcher, L.M. and Clarke, L.P. (1998) Review and evaluation of MRI nonuniformity corrections for brain tumor response measurements. *Medical Physics*, Vol. 25, 1655–1666.

Venkatesan, R., Lin, W. and Haacke, E.M. (1998) Accurate determination of spin-density and T_1 in the presence of RF field inhomogeneities and flip-angle miscalibration. *Magnetic Resonance in Medicine*, Vol. 40, 592–602.

Vescovo, E., Levick, A., Childs, C., Machin, G., Zhao, S. and Williams, S.R. (2013) High-precision calibration of MRS thermometry using validated temperature standards: effects of ionic strength and protein content on the calibration. *NMR in Biomedicine*, Vol. 26, 213–223.

Wagnerova, D., Herynek, V., Malucelli, A., Dezortova, M., Vymazal, J., Urgosik, D., Syrucek, M., Jiru, F., Skoch, A., Bartos, R., Sames, M. and Hajek, M. (2012) Quantitative MR imaging and spectroscopy of brain tumours: a step forward? *European Radiology*, Vol. 22, 2307–2318.

Walsh, A.J., Blevins, G., Lebel, R.M., Seres, P., Emery, D.J. and Wilman, A.H. (2014) Longitudinal MR imaging of iron in multiple sclerosis: an imaging marker of disease. *Radiology*, Vol. 270, 186–196.

Walsh, D.O., Gmitro, A.F. and Marcellin, M.W. (2000) Adaptive reconstruction of phased array MR imagery. *Magnetic Resonance in Medicine*, Vol. 43, 682–690.

Wang, D., Doddrell, D.M. and Cowin, G. (2004) A novel phantom and method for comprehensive 3-dimensional measurement and correction of geometric distortion in magnetic resonance imaging. *Magnetic Resonance Imaging*, Vol. 22, 529–542.

Wang, Z.J., Chia, J.M. and Rollins, N.K. (2011) A quality assurance protocol for diffusion tensor imaging using the phantom from American College of Radiology. *Medical Physics*, Vol. 38, 4415–4421.

Wanspura, J.P., Holland, S.K., Dunn, R.S. and Ball, W.S. Jr (1999) NMR relaxation times in the human brain at 3.0 tesla. *Journal of Magnetic Resonance Imaging*, Vol. 9, 531–538.

Weiger, M., Pruessmann, K.P. and Boesiger, P. (2002) 2D SENSE for faster 3D MRI. *MAGMA*, Vol. 14, 10–19.

Weisskoff, R.M. (1996) Simple measurement of scanner stability for functional NMR imaging of activation in the brain. *Magnetic Resonance in Medicine*, Vol. 36, 643–645.

Wheeler, D.J. (1992) *Understanding Statistical Process Control,* 2nd edition. Longman Higher Education, Knoxville, TN, USA.

Winfield, J.M., Douglas, N.H., de Souza, N.M. and Collins, D.J. (2014) Phantom for assessment of fat suppression in large field-of-view diffusion-weighted magnetic resonance imaging. *Physics in Medicine and Biology*, Vol. 59, 2235–2248.

Wolfram (2015) *Mathematica.* Available at: http://www.wolfram.com/mathematica/ [accessed on 4/06/2015].

Woo, D.C., Kim, B.S., Jung, S.L., Park, H.J., Rhim, H.S., Jahng, G.H. and Choe, B.Y. (2007) Development of a cone-shape phantom for multi-voxel MR spectroscopy. *Journal of Neuroscience Methods*, Vol. 162, 101–107.

Wood, J.C. (2007) Magnetic resonance imaging measurement of iron overload. *Current Opinion in Hematology*, Vol. 14, 183–190.

Woodward, P.K., Bluemke, D.A., Cascade, P.N., Finn, J.P., Stillman, A.E., Higgins, C.B., White, R.D. and Yucel, E.K. (2006) American College of Radiology. ACR practice guideline for the performance and interpretation of cardiac magnetic resonance imaging (MRI). *Journal of the American College of Radiology: JACR*, Vol. 3, 665–676.

Wu, T. and Felmlee, J.P. (2002) A quality control program for MR-guided focused ultrasound ablation therapy. *Journal of Applied Clinical Medical Physics Research A*, Vol. 3, 162–167.

Xin, X., Han, J., Wang, D., Feng, Y., Feng, Q. and Chen, W. (2012) Development of a calibration phantom set for MRI temperature imaging system quality assurance. *Academic Radiology* 19, 740–745.

Xinapse (2016) Jim 7 Software; Xinapse Systems. Available at: http://www.xinapse.com/j-im-7-software/ [accessed on 8/06/2016].

Yablonskiy, D.A., Neil, J.J., Raichle, M.E. and Ackerman, J.J. (1998) Homonuclear J coupling effects in volume localized NMR spectroscopy: pitfalls and solutions. *Magnetic Resonance in Medicine*, Vol. 39, 169–178.

Yang, X., Sammet, S., Schmalbrock, P. and Knopp, M.V. (2010) Postprocessing correction for distortions in T_2* decay caused by quadratic cross-slice B_0 inhomogeneity. *Magnetic Resonance in Medicine*, Vol. 63, 1258–1268.

Yoshimura, K., Kato, H., Kuroda, M., Yoshida, A., Hanamoto, K., Tanaka, A., Tsunoda, M., Kanazawa, S., Shibuya, K., Kawasaki, S. and Hiraki, Y. (2003) Development of a tissue-equivalent MRI phantom using carrageenan gel. *Magnetic Resonance in Medicine*, Vol. 50, 1011–1017.

Zhou, X.J. (2004) Correction gradients: Concomitant-field correction gradients. In: *Handbook of MRI Pulse Sequences*, edited by M.A. Bernstein, K.F. King and X.J. Xhou. Elsevier Academic Press, Burlington, MA, USA.

Zhou, X.J., Du, Y.P., Bernstein, M.A., Reynolds, H.G., Maier, J.K. and Polzin, J.A. (1998a) Concomitant magnetic-field-induced artifacts in axial echo planar imaging. *Magnetic Resonance in Medicine*, Vol. 39, 596–605.

Zhou, X.J., Tan, S.G. and Bernstein, M.A. (1998b) Artifacts induced by concomitant magnetic field in fast spin-echo imaging. *Magnetic Resonance in Medicine*, Vol. 40, 582–591.

Zhu, T., Hu, R., Taylor, M., Tso, Y., Yiannoutsos, C., Navia, B., Mori, S., Ekholm, S., Schifitto, G. and Zhong, J. (2011) Quantification of accuracy and precision of multicenter DTI measurements: A diffusion phantom and human brain study. *Neuroimage*, Vol. 56, 1398–1411.

Zhuo, J. and Gullapalli, R.P. (2006) AAPM/RSNA physics tutorial for residents: MR artifacts, safety, and quality control. *Radiographics*, Vol. 26(1), 275–297.

Index

Page numbers for index words present in headings are shown in boldface.

Page numbers for index words present in table titles, tables, figure captions or within figures themselves are shown in italic.

www.ingramcontent.com/pod-product-compliance
Ingram Content Group UK Ltd.
Pitfield, Milton Keynes, MK11 3LW, UK
UKHW051335170926
15239UKWH00010B/26

9 781903 613610